Fundamental Electrical Experiments

电子学基础实验

方 元 编 著

南京大学出版社

图书在版编目(CIP)数据

电子学基础实验 = Fundamental Electrical Experiments：英文、汉文 / 方元编著. — 南京 ：南京大学出版社，2023.1
ISBN 978-7-305-26251-7

Ⅰ. ①电… Ⅱ. ①方… Ⅲ. ①电子学－实验－高等学校－教学参考资料－英、汉 Ⅳ. ①TN01-33

中国版本图书馆 CIP 数据核字(2022)第 213382 号

出版发行　南京大学出版社
社　　址　南京市汉口路 22 号　　　　邮　编　210093
出 版 人　金鑫荣
书　　名　**电子学基础实验**
　　　　　Fundamental Electrical Experiments
编　　著　方 元
责任编辑　吕家慧　　　　　　　编辑热线　025-83597482
照　　排　南京南琳图文制作有限公司
印　　刷　南京京新印刷有限公司
开　　本　787 mm×1092 mm　1/16　印张 9.75　字数 220 千
版　　次　2023 年 1 月第 1 版　2023 年 1 月第 1 次印刷
ISBN 978-7-305-26251-7
定　　价　39.00 元

网址：http://www.njupco.com
官方微博：http://weibo.com/njupco
微信服务号：njuyuexue
销售咨询热线：(025) 83594756

前　言

电路分析实验和模拟电路实验一直是南京大学电子信息学科重要的基础实验课程.为适应我校课程体系和教学改革的要求,我们编写了这本基础电子学实验教材.

本教材主要涵盖四个方面的实验内容.

1. 基本电子测量仪器的使用.测量仪器在现代电子设计研发中发挥着重要的作用.测量仪器贯穿在整个实验课程中.在学习中,除了学会熟练操作,还应了解各种仪器的性能、特性和它们的局限性,以便在测量中获得可信的结果.

2. 电路分析实验.由于课时的限制,我们在这门课程中重点设计了基尔霍夫定律和等效电路的验证实验,这也是电路分析理论中比较重要的内容.灵活运用这些方法,将有益于对后续实验的分析.

3. 晶体管放大电路.由于电子技术的发展,现在已经很少用晶体管直接搭建放大器了,甚至将晶体管用作线性元件的应用场合也不多,更多的是使用集成运算放大器.而集成运算放大器则是以晶体管放大器为基础,并且晶体管放大器目前仍然是模拟电路教学的重要内容之一.本课程安排了两个晶体管放大电路实验,它们基本涵盖了放大电路的主要参数概念和测量方法.

4. 集成运算放大器.集成运算放大器应用及其广泛.本课程重点研究了放大器、滤波器、振荡器实验,并借此为学生建立一个电路系统的频率响应概念.

在基础学习阶段,大多数实验都是验证性的.验证性实验是基础教学的重要组成部分.它除了帮助学生理解和巩固基础理论知识,也是培养学生科学研究的方法,提高科学素质的途径.实验重在过程.实验结果的正确性当然是评估实验能力的一方面,但更重要的是,要学会分析测量数据的可靠性,学习科学地描述实验现象、归纳实验结果、总结规律,结合理论知识阐述结果的合理性.只有这样,才算是真正掌握了理论知识.

除了通过理论分析与实验结果相印证,在如今的电子系统的分析与设计中,还普遍使用了大量的工具软件,特别是仿真软件.这些仿真软件由专业的工程师设计,根据理论模型,通过严格的数学计算,将电路的分析结果展现给使用者.教材中对电路仿真软件 NGSPICE 做了初步的介绍.优秀的仿真工具,比经过简化的理论分析结果更接近真实.仿真实验不受实验室条件和开放时间的限制.建议学生在实验环节积极动手操作的同时,在实验预习阶段,除了进行初步的理论分析,也学习掌握仿真工具的使用,通过仿真,对实验结果有一个大致的预期.这样,在实验中发现与预期结果的偏差时,可以及时发现问题,及时与指导教师交流,提高课堂效率.

教材中有意融合了信号分析的理论,并通过一个综合性实验帮助学生理解相关概念以及它们在电路系统中的地位,为今后学习信号与系统理论做好思想准备.

教材中的大多数实验按 4 个学时的工作量设计,课程总计 64 学时.晶体管放大器实验、运算放大器入门实验和综合实验"信号分解与合成",因涉及的测量量比较多,实验指导教师可根据学生水平和课程要求适当增补实验学时.特别是晶体管放大器实验,考虑到实验的完整性,教材将单级无反馈放大器和多级反馈放大电路作为一个整体进行设计.在实际教学实践中,可以考虑对其进行适当地分解.

电子技术的发展日新月异,基础电子实验的教学也应不断地完善和提高.限于笔者的知识水平和认知能力,书中难免存在各种不足甚至错误,恳请读者批评指正.

作者

2022 年夏

Contents

◀◀◀

1

Electronic Measuring Equipments
电子测量仪器

In all of the electrical experiments, you will be using various types of test equipment to generate or to measure aspects of electricity you cannot directly create by hand or feel by our bodies.

1.1 Multimeter (万用表)

A multimeter is a measuring instrument that can measure multiple electrical properties. A typical multimeter can measure voltage, resistance, current and other electrical quantities.

Analog multimeters use a microammeter with a moving pointer to display readings. Digital multimeters have numeric displays and have made analog multimeters obsolete as they are cheaper, more precise, and more physically robust than analog multimeters (Figure 1.1).

1.1.1 Function Description (功能描述)

Figure 1.2 is the front panel of digital multimeter DM3068. The multimeter has the following functions.

1. USB Host. The system configuration or measurement data can be stored into USB storage device and be recalled when required.
2. LCD. The LCD panel can display the current function menus, measurement parameter settings, system status, prompt messages and so on.

▷▷▷ ──

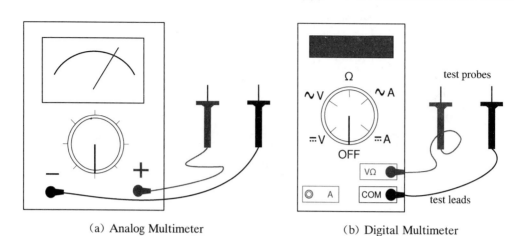

(a) Analog Multimeter   (b) Digital Multimeter

Figure 1.1　Analog and Digital Multimeters

图 1.1　模拟和数字万用表

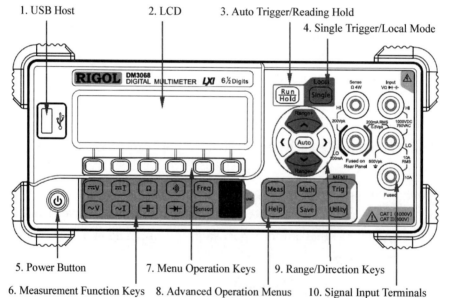

1. USB Host　　2. LCD　　3. Auto Trigger/Reading Hold

4. Single Trigger/Local Mode

5. Power Button　　7. Menu Operation Keys　　9. Range/Direction Keys

6. Measurement Function Keys　　8. Advanced Operation Menus　　10. Signal Input Terminals

Figure 1.2　Multimeter DM3068 Front Panel Overview

图 1.2　DM3068 数字万用表前面板一览

3. Auto Trigger/Reading Hold. Switch between auto trigger and reading hold functions.

4. Single Trigger/Local Mode. Switch between single trigger and local mode.

5. Power Button. Turn on or off the multimeter. Use $\boxed{\text{Utility}}$→System→Cfg→ Switch and select "ON" or "OFF" to enable or disable this key.

6. Measurement Function Keys.

   Left to right in first row:

   • DC (Direct Current) Voltage Measurement (DCV).

- DC Current Measurement（DCI）.
- Resistance Measurement（OHM）.
- Continuity Test（CONT）.
- Freq Frequency/Period Measurements（FREQ/PERIOD）.
- Preset Quickly save or recall 10 groups of instrument settings.

Left to right in the second row：

- AC（Alternating Current）Voltage Measurement（ACV）.
- AC Current Measurement（ACI）.
- Capacitance Measurement（CAP）.
- Diode Test（DIODE）.
- Any Sensor Measurements（SENSOR）, such as DCV, DCI, 2WR, 4WR, FREQ, TC （thermoelectric couple）, RTD （resistance temperature detectors）and THERM（thermistor）.
- Secondary function key.

7. Menu Operation Keys. Press any softkey to activate the corresponding menu.

8. Advanced Operation Menus.
- Meas Set all the measurement parameters.
- Math Perform math operations（statistic, P/F, dBm, dB, REL）on the measurement results.
- Trig Provide auto, single, external and level trigger.
- Save Save the system configuration and measurement data into internal memory or external USB storage device.
- Utility Select command set for the multimeter; configure the interface parameters and system parameters; execute self-test and view the system information and error messages.
- Help Provide help information for common operations and the use method of the built-in help.

9. Range/Direction Keys. Configure the measurement parameters, increase or decrease the measurement range, or set auto range, etc.

10. Signal Input Terminals. The measured signal （device）will be connected into the multimeter through these terminals. Different measurement objects have different connection methods.

1.1.2　Terminals（连接端口）

DM3068 is designed with many measurement functions. After selecting the

▶▶▶

desired measurement function，please connect the signal （device） under test according to the method shown in Figure 1.3.

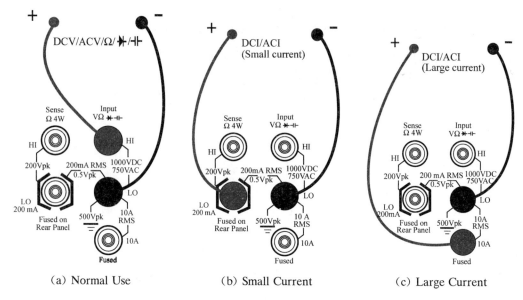

(a) Normal Use　　　　(b) Small Current　　　　(c) Large Current

Figure 1.3　Measurement Connection

图 1.3　测量探头连接方式

⚠️ In order to protect the multimeter，please execute DC/AC current measurement following the requirements below：

• Do not connect the 10 A and LO Sense/200 mA input terminals into the current measuring circuit at the same time.

• Select a proper current input terminal according to the estimated current magnitude before connecting the multimeter to AC power supply if you want to use current measurement.

• Only use 10 A and LO terminals for measurements when the AC + DC RMS value of the current under measurement goes within 200 mA and 10 A.

For using of connectors in rear panel and more information，please refer to the DM3068 user guide.

1.2　Oscilloscope （示波器）

An oscilloscope is a type of electronic test instrument that graphically displays varying signal voltages，usually as a calibrated two-dimensional plot of one or more signals as a function of time. The displayed waveform can then be analyzed for properties such as amplitude, frequency, rise time, time interval, **distortion**, and others.

1.2.1　Digital Storage Oscilloscope（数字存储示波器）

Early oscilloscopes used **cathode ray tubes**（CRTs）as their display element（hence they were commonly referred to as CROs）and linear amplifiers for signal processing. CROs were later largely superseded by **digital storage oscilloscopes**（DSOs）with thin panel displays, fast **analog-to-digital converters**（ADC）and digital signal processors. General principle of DSO is shown in Figure 1.4.

The measured physical quantity is converted into voltage first. The processor then reads the value converted by the analog-to-digital converter, stores the data in memory, calculates and displays the results on the screen.

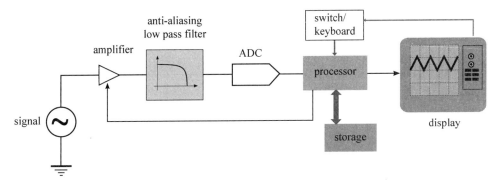

Figure 1.4　Schematic Principle of Digital Storage Oscilloscope

图 1.4　数字存储示波器原理框图

Digital oscilloscope can be adjusted so that repetitive signals can be observed as a persistent waveform on the screen. A storage oscilloscope can capture a single event and display it continuously, so the user can observe events that would otherwise appear too briefly to see directly.

1.2.2　Function of Keysight 1102A（是德 1102A 型示波器功能说明）

The front panel of Keysight 1000X-Series oscilloscope is shown in Figure 1.5. The function of the keys and knobs on the panel is listed below：

1. Power switch, to switch between "ON" and "OFF".
2. Softkeys. The functions of these keys change based upon the menus shown on the display next to the keys.

   The $\boxed{\text{Back}}$ key moves back in the softkey menu hierarchy. At the top of the hierarchy, the $\boxed{\text{Back}}$ key turns the menus off, and oscilloscope information is shown instead.

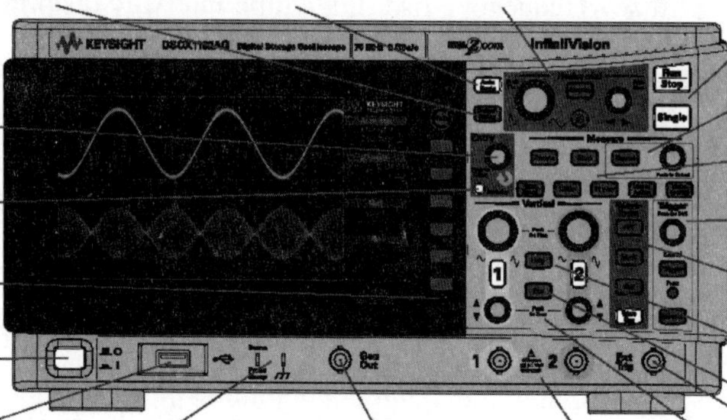

5. [Default Setup] key   6. [Auto Scale] key   7. Horizontal and Acquisition controls

4. Entry knob

3. [Intensity] key

2. Softkeys

1. Power switch

8. Run Control keys

9. Measure controls

10. Tools keys

11. Trigger controls

12. Waveform keys

13. [Help] key

14. [Bus] key

20. USB Host prot   19. Demo/Probe Comp and Ground terminals   18. Waveform generator output   17. Analog channel inputs   16. Vertical controls   15. Ext Trig input

Figure 1.5   Oscilloscope Keysight 1000X‑Series Front Panel

图 1.5   Keysight 1000X 系列示波器面板

3. Intensity, to change the intensity of drawings combining with Entry knob.

4. Entry knob. The Entry knob is used to select items from menus and to change values.

5. Default Setup, to restore the oscilloscope's default settings.

6. Auto Scale. When pressed, the oscilloscope will quickly determine which channels have activity, and it will turn these channels on and scale them to display the input signals.

7. Horizontal and Acquisition controls. The Horizontal and Acquisition controls consist of:

- Horizontal scale knob. Turn the knob in the horizontal section that is marked to adjust the time/div setting. Push the knob to toggle between fine and coarse adjustment.

- Horizontal position knob. Turn the knob marked to pan through the waveform data horizontally. By turning the knob clockwise or counterclockwise, you can see the captured waveform before the trigger or after the trigger(turn the knob counterclockwise).

- Acquire . Press this key to open the Acquire menu where you can select the Normal, XY, and Roll time modes, enable or disable Zoom, and select the trigger time reference point.

- Zoom key, to split the oscilloscope display into Normal and Zoom sections without opening the Acquire menu.

8. Run Control keys. When the Run/Stop key is green, the oscilloscope is

running, that is, acquiring data when trigger conditions are met. When the Run/Stop key is red, data acquisition is stopped. Push this key to switch between "RUN" and "STOP" mode.

To capture and display a single acquisition (whether the oscilloscope is running or stopped), press Single . The Single key is yellow until the oscilloscope triggers.

9. Measure controls. The Measure controls consist of:
   - Analyze , to access analysis features like trigger level setting, measurement threshold setting, video trigger automatic set up and display, or digital voltmeter.
   - Meas , to access a set of predefined measurements.
   - Cursors , to open a menu that lets you select the cursors mode and source.
   - Cursors knob, to select cursors from a popup menu.

10. Tools keys. The Tools keys consist of:
   - Save/Recall , to save oscilloscope setups, screen images, waveform data, or mask files or to recall setups, mask files or reference waveforms.
   - Utility , to access the Utility menu, which lets you configure the oscilloscope's I/O settings, use the file explorer, set preferences, access the service menu, and choose other options.
   - Display , to access the menu where you can enable persistence, adjust the display grid intensity, label waveforms, add an annotation, and clear the display.
   - Quick Action , to perform the selected quick action: measure all snapshot, print, save, recall, freeze display, and more.
   - Save to USB , to perform a quick save to a USB storage device.

11. Trigger controls. The Trigger controls determine how the oscilloscope triggers to capture data. These controls consist of:
   - Level knob, to adjust the trigger level for a selected analog channel. Push the knob to set the level to the waveform's 50% value. If AC coupling is used, pushing the Level knob sets the trigger level to about 0 V. The position of the trigger level for the analog channel is indicated by the trigger level icon (if the analog channel is on) at the far left side of the display. The value of the analog channel trigger level is displayed in the upper-right corner of the display.

▶▶▶

- Trigger , to select the trigger type (edge, pulse width, video, etc.).
- Force, causes a trigger (on anything) and displays the acquisition.
- External , to set external trigger input options.

12. Waveform keys. The additional waveform controls consist of:
   - FFT , provides access to FFT spectrum analysis function.
   - Math , provides access to math (add, subtract, etc.) waveform functions.
   - Ref , provides access to reference waveform functions. Waveforms that are saved can be displayed and compared against other analog channel or math waveforms.
   - Wave gen , to access waveform generator functions (On G-suffix models).

13. Help , opens the Help menu where you can display overview help topics and select the Language.

14. Bus , opens the Bus menu. Either display a bus made up of the analog channel inputs and the external trigger input where channel 1 is the least significant bit and the external trigger input is the most significant bit, or enable serial bus decodes.

15. Ext Trig input. External trigger input BNC① connector.

16. Vertical controls. The Vertical controls consist of:
   - Analog channel on/off keys, to switch a channel on or off, or to access a channel's menu in the softkeys.
   - Vertical scale knob, to change the vertical sensitivity (gain) of each analog channel. Push the channel's vertical scale knob to toggle between fine and coarse adjustment.
   - Vertical position knobs, to change a channel's Vertical position on the display.

17. Analog channel inputs. Attach oscilloscope probes or BNC cables to these BNC connectors. In the 1000X-Series oscilloscopes, the analog channel inputs have 1 MΩ impedance, and there is no automatic probe detection, so you must properly set the probe attenuation for accurate measurement results.

18. Waveform generator output. On G-suffix models, the built-in waveform

---

① BNC (Bayonet Neill-Concelman): The BNC connector is a miniature quick connect/disconnect electronic equipment connector used for coaxial cable. It was named after its inventor, Paul Neill of Bell Labs and Carl Concelman of Amphenol Corp.

generator can output sine，square，ramp，pulse，DC，or noise on the Gen Out BNC.

19. Demo/Probe Comp，Ground terminals. Demo terminal outputs the Probe Comp signal which helps you match a probe's input capacitance to the oscilloscope channel to which it is connected.

20. USB Host port. Connect a USB compliant mass storage device (flash drive，disk drive，etc.) to save or recall oscilloscope setup files and reference waveforms or to save data and screen images，or connect a USB compliant printer.

⚠ Do not connect a host computer to the oscilloscope's USB host port. A host computer sees the oscilloscope as a device. When needed，connect the host computer to the oscilloscope's device port (on the rear panel).

1.2.3　Oscilloscope Probe（示波器探头）

Oscilloscopes do not normally use simple wires ("flying leads") to connect to the device under test. Instead，a specific scope probe is used，which uses a **coaxial cable** to transmit the signal from the tip of the probe to the oscilloscope (Figure 1.6). This cable has two main benefits： it protects the signal from external electromagnetic interference，improving accuracy for low-level signals；and it has a lower inductance than flying leads，making the probe more accurate for high-frequency signals.

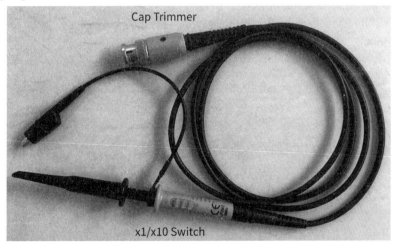

Figure 1.6　Oscilloscope Probe with Calibrating Trimmer and Attenuation Select Switch

图 1.6　带有校准微调和衰减切换的示波器探头

Although coaxial cable has lower inductance than flying leads, it has higher capacitance: a typical 50 Ω cable has about 90 pF per meter. Consequently, an one-meter high-impedance direct coaxial probe may load the circuit with a capacitance of about 110 pF and a resistance of 1 MΩ. To minimize loading, attenuator probes (e. g. ×10 probes) are used. The ×10 scope probe uses a series resistor (9 MΩ) and a capacitor for compensation (see Figure 1.7).

Assume the output resistance of the test signal source R_S is 1 kΩ. When the probe is switched to ×1 (C_P and R_P short circuited), the signal entering the oscilloscope (node A) is

$$V_A(j\omega) = \frac{R \parallel \dfrac{1}{j\omega C}}{R_S + \left(R \parallel \dfrac{1}{j\omega C} \right)} V_S(j\omega) \tag{1.1}$$

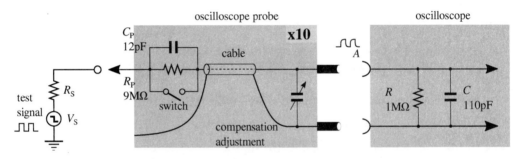

Figure 1.7 Oscilloscope Probe Circuit Internals

图 1.7 示波器探头内部结构

For $R = 1$ MΩ, $C = 110$ pF, $|V_A| = 0.999|V_S|$ at 1 kHz, but $|V_A| = 0.586|V_S|$ at 2 MHz. High frequencies are greatly attenuated.

With compensation circuit switched on (×10), the frequency response at point A should be

$$V_A(j\omega) = \frac{R \parallel \dfrac{1}{j\omega C}}{R_S + \left(R_P \parallel \dfrac{1}{j\omega C_P} \right) + \left(R \parallel \dfrac{1}{j\omega C} \right)} V_S(j\omega) \tag{1.2}$$

Although there is attenuation at low frequencies ($|V_A| \approx 0.1|V_S|$ at 1 kHz), the frequency response of the system maintains almost the same in a wider frequency range ($|V_A| \approx 0.08|V_S|$ at frequency 10 MHz). ×10 probe has a much higher bandwidth than the ×1 scope probe. Figure 1.8 shows the frequency responses at different probe switches.

An improperly compensated oscilloscope probe will result in inaccurate high-frequency measurements. Therefore it's better verifying and correcting the probe compensation when first connecting a probe to an oscilloscope. Figure 1.9 shows the square waves displayed on the screen from an undercompensated, properly compensated and overcompensated oscilloscope probe.

Most oscilloscope probes can be switched between $\times 1$ and $\times 10$. As the $\times 10$ probe attenuates the signal by a factor of ten, the signal entering the scope itself will be reduced. This has to be taken into account. Some oscilloscopes automatically adjust the scales according to the probe present, although not all are able to do this. It is worth checking before reading on the oscilloscope.

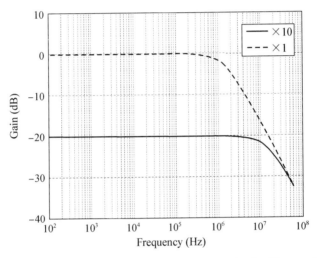

Figure 1.8　Frequency Response of DSO Affected by Oscilloscope Probe

图 1.8　DSO 探头对频率响应的影响

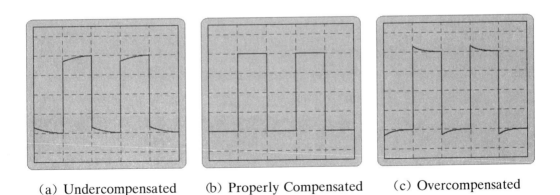

(a) Undercompensated　　(b) Properly Compensated　　(c) Overcompensated

Figure 1.9　A Square Wave from a Compensated Oscilloscope Probe

图 1.9　通过示波器探头补偿的方波信号

1.3　Function Generator（函数发生器）

A function generator is usually a piece of electronic test equipment or software used to generate different types of electrical waveforms over a wide range of frequencies. In addition to producing sine waves, function generators may typically produce other repetitive waveforms including sawtooth and triangular waveforms, square waves, and pulses, etc.

Another feature included on many function generators is the ability to add a DC offset. Figure 1.11 shows the diagram of a digital function generator. The processor converts the input parameters to control signals to the DDS (Direct Digital Synthesis) to generate desired waveform, send the signal through DAC (Digital to Analog Converter) and output analog signal.

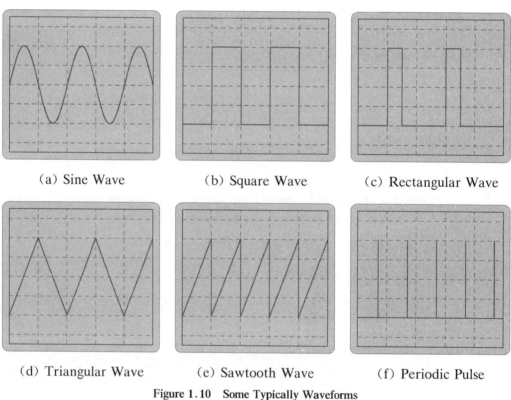

(a) Sine Wave　　　(b) Square Wave　　　(c) Rectangular Wave

(d) Triangular Wave　　　(e) Sawtooth Wave　　　(f) Periodic Pulse

Figure 1.10　Some Typically Waveforms

图 1.10　一些典型的波形

RIGOL's DG4162 is a dual-channel arbitrary waveform generator. Figure 1.12 shows its front panel. The brief usage is summarized below:

　1. Power Key, to turn the generator on or off. When the power key is turned off, the generator is in standby mode. The generator is in power-off mode

only when the power cable at the rear panel is pulled out.

2. USB Host. Support USB storage device or other RIGOL equipments interconnection. When a USB storage is plugged in, the device can read the waveform files or state files in the USB storage device, store the current instrument state or edited arbitrary waveform data into the USB storage device, or store the content currently displayed on the screen in the specified picture format (BMP or JPEG).

3. Menu Softkey.

4. Page Up/Down. Open the previous or next page of the current function menu.

5. CH1 Output. BNC connector with 50 Ω nominal output impedance. When Output1 is enabled (the backlight turns on), this connector output waveform according to the current configuration of CH1.

6. CH1 Sync Output. BNC connector with 50 Ω nominal output impedance. When the sync output of CH1 is enabled, this connector outputs the sync signal corresponding to the current settings of CH1.

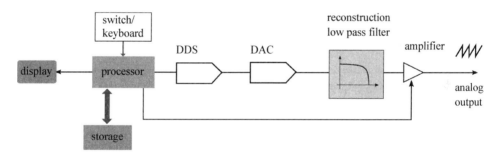

Figure 1.11　A Diagram of Function Generator

图 1.11　函数信号发生器框图

7. Channel Control Area.

CH1, CH2: to select the corresponding channel to output.

Trigger1, Tirgger2: to trigger the corresponding channel to generate a sweep or burst output manually in sweep or burst mode.

CH1⇌CH2: execute channel copy.

8. CH2 Output (see CH1 Output).

9. CH2 Sync Output (see CH1 Sync Output).

10. Counter. When the counter is turned on, the backlight of the key is illuminated and the left indicator flickers.

11. Numeric Keyboard, used to input parameters.

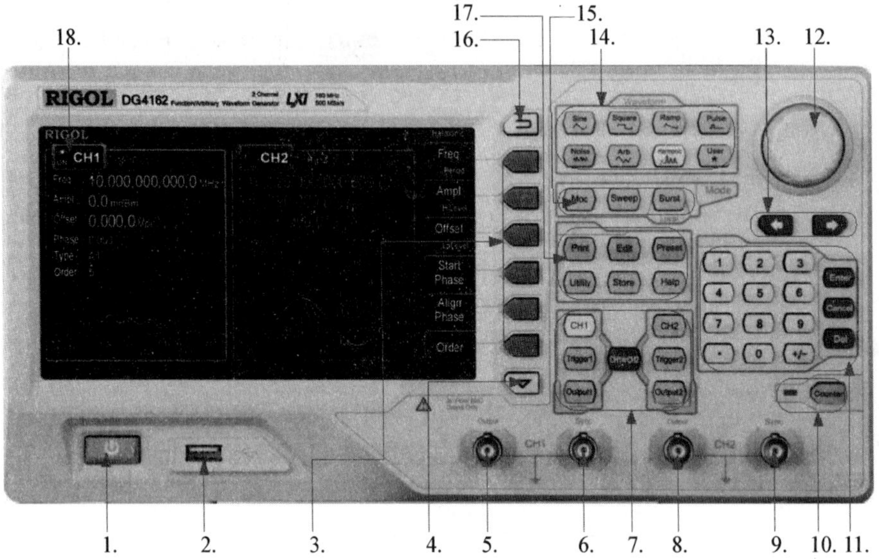

▶ ▶ ▶

Figure 1.12　Function Generator RIGOL DG4162

图 1.12　RIGOL DG4162 信号发生器

12. Knob. During parameter setting, it is used to increase (clockwise) or decrease (counterclockwise) the current highlighted number (associate with ⇐ and ⇒ keys), to select file storage location or to select the file to be recalled when storing or recalling file, to switch the character in the soft keyboard when entering filename, or to select built-in waveform when defining the shortcut waveform of User .

13. ⇐ , ⇒ . See above.

14. Waveform Selection Area.

15. Modes.

 Mod : generate the modulated waveforms.

 Sweep : generate the frequency sweeping signal of sine, square, ramp and arbitrary waveforms.

 Burst : generate burst waveforms of sine, square, ramp, pulse and arbitrary waveform.

16. Return. This key is used to return to the previous menu.

17. Shortcuts/Utility.

18. LCD screen is used to display the current function menu and parameters setting, system state as well as prompt messages.

Overvoltage protection of the output channel will take effect once any of the following conditions is met:

- Amplitude setting in the generator is greater than 4 V_{PP}; the input voltage is greater than ± 11.25 V (± 0.1 V) and frequency is lower than 10 kHz.
- Amplitude setting in the generator is lower than or equal to 4 V_{PP}; the input voltage is greater than ± 4.5 V (± 0.1 V) and frequency is lower than 10 kHz.
- The message "Overload protect, The output is off!" will appear on the screen when overvoltage protection takes effect.

For connectors in rear panel, please refer to the DG4000 manual and user's guide.

1.4 Objective (实验目的)

The purpose of this lab is to introduce to the students the commonly used electronic devices and their characteristics.

1.5 Experimental Procedure (实验步骤)

1.5.1 Generating and Measuring Sine Wave (正弦波产生与测量)

The function generator outputs sine wave (1 V_{PP}, offset = 100 mV) in different frequencies (till maximum frequency the generator can output). Read the frequencies and magnitudes using both multimeter and oscilloscope, copy the readings and fill them in Table 1.1.

The magnitude of an AC signal usually measured in terms of amplitude, peak-to-peak or RMS (root mean square), see Figure 1.13.

Peak-to-peak voltage is the distance from the lowest negative amplitude, or trough, to the highest positive amplitude, or crest, of the AC voltage waveform. In other words, peak-to-peak voltage denotes the full height of the waveform.

RMS, or root mean square voltage (also known as effective voltage), is a method of denoting an AC waveform as an equivalent voltage which represents the DC voltage value that will produce the same power dissipation. In other words, the RMS value allows an AC waveform to be specified as a DC, because it is the equivalent DC voltage that delivers the same amount of power to a load in a circuit as the AC signal does over its cycle. The RMS voltage of a periodic signal $v(t)$ is calculated by

$$V_{RMS} = \sqrt{\frac{1}{T}\int_0^T v^2(t)\,dt} \tag{1.3}$$

A sinusoidal signal of 1 V amplitude with offset = 0 V is equivalent to 2 V_{PP} or 0.707 V_{RMS}. You should always notice what scales you are reading.

1.5.2 Measurement Affected by the Probe (探头对测量的影响)

The function generator generates square waves with different frequencies (1 V_{PP}, 50% duty cycle, offset = 0). Use oscilloscope to watch the waveform, read the amplitude from the oscilloscope. If necessary, switch the oscilloscope probe between ×10 and ×1. Fill your recorded data in Table 1.2. In what circumstances will the ×10 switch be useful?

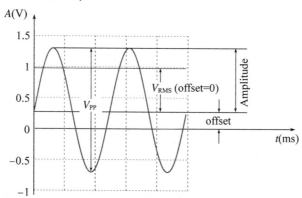

Figure 1.13　Amplitude, V_{PP} and V_{RMS}

图 1.13　振幅、峰—峰值和均方根

Table 1.1　Sine Wave of Different Frequencies(1 V_{PP})

表 1.1　不同频率的正弦波(1 V_{PP})

Generator Frequency (Hz)	Frequency (Hz)		Amplitude (V)		Remark
	Multimeter	Oscilloscope	Multimeter	Oscilloscope	
20					
200					
2 k					
20 k					
200 k					
2 M					
20 M					
(max.)					

1.5.3 Sine Wave of Different Magnitudes (信号幅度测量)

The function generator generates an 1 kHz sine wave of different amplitudes. Measure the amplitudes using multimeter and oscilloscope. Write down your record in Table 1.3. Note how you read values on the oscilloscope, especially at a low **signal-to-noise ratio**.

Table 1.2 Square Wave of Different Frequencies (1 V_{PP})

表 1.2 不同频率的方波(1 V_{PP})

Frequency (Hz)	Amplitude (V)	Waveform Description	Probe Present $\times 1/ \times 10$
100			
10 k			
1 M			
10 M			
(max.)			

Table 1.3 Sine Wave of Different Amplitudes (1 kHz)

表 1.3 不同幅度的正弦波(1 kHz)

Amplitude (mV)	Multimeter (mV)	Oscilloscope (mV)	Remark
5000			
500			
50			
5			
(min.)			

1.5.4 DC Coupling and AC Coupling (直流耦合与交流耦合)

The function generator generates square waves (1 V_{PP}, offset = 200 mV) with different frequencies. Push the channel select button of the oscilloscope, switch the coupling mode between AC and DC from the popup menu, watch the waveforms shown on the screen and sketch them in Table 1.4.

1.5.5 Harmonic Analysis (谐波分析)

According to Fourier[①] analysis theory, a periodic signal $s(t)$ can be written as

---

① Jean-Baptiste Joseph Fourier (Mar. 21, 1768 – May 16, 1830), French mathematician and physicist.

a weighted summation of a series of sine waves.

$$s(t) = a_0 + \sum_{n=1}^{\infty} a_n \sin(2\pi n f_0 t + \phi_n)$$

where a_0 is DC offset，a_1 is the amplitude of fundamental sine wave，f_0 is **fundamental frequency**，and a_n ($n > 1$) is the amplitude of the n-th harmonic frequency.

Table 1.4　Observed Waveforms Using DC or AC Coupling

表 1.4　使用直流或交流耦合观察到的波形

	10 Hz	1 kHz
AC coupling		
DC coupling		

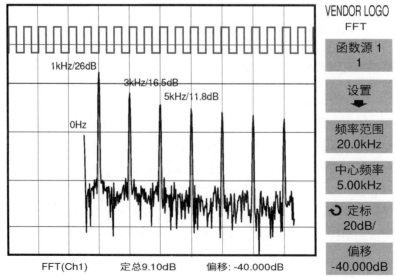

Figure 1.14　FFT Function of Oscilloscope

图 1.14　示波器的 FFT 功能

The generator outputs a rectangular wave of different duty cycles. Use the FFT function of oscilloscope to display spectrum of the signal. Record the fundamental, second and third harmonic amplitudes L_1, L_2 and L_3 in dB scale.

Table 1.5　Harmonic Analysis (1 kHz, 1 V$_{PP}$)

表 1.5　谐波分析(1 kHZ, 1 V$_{PP}$)

Duty cycle	L_1(dB)	L_2(dB)	L_3(dB)	d_2(%)	d_3(%)
20%					
50%					
80%					

When harmonic magnitudes are measured in dB, the n-th harmonic component related to fundamental sine wave is calculated by

$$d_n = 10^{\frac{L_n - L_1}{20}} \times 100\% \tag{1.4}$$

1.6　Questions(问题)

Try to answer the following questions：

1. What is RMS voltage of a triangular wave with 1 V$_{PP}$?

2. To measure a low frequency square wave using oscilloscope with AC coupling or DC coupling，why do they look different?

3. You have multimeters and oscilloscopes by hand. Which device will you choose to measure signal amplitude with the frequency of mega hertz or kilo hertz? How do you measure signal with the amplitude of several millivolts or hundred millivolts? Please explain your reason.

［实验简介］

万用表、示波器、信号发生器和电源常被称为基础电子学实验的"四大件".了解它们的特性、熟练掌握这些仪器设备的使用方法,是之后顺利进行其他实验的基础.

本实验重点设计了使用示波器、万用表结合信号发生器测量电压的几个实验.通过这些实验,学习如何产生正确的测量信号,如何从仪表中读取反映信号特征的数据.读者应从中了解:各种仪器都有其工作范围,有其自身的局限性.特别对于在高频、微弱以及低信噪比测量场合,应尽可能使用多种分析手段,这样才能获得正确的测量结果.

［Vocabulary］

multimeter：多功能表,万用电表

microammeter：微安计

analog：模拟的

digital：数字的

LCD：液晶显示器（Liquid Crystal Display）

trigger：触发

thermoelectric couple：热电偶

input terminal：输入端

digital storage oscilloscope：数字存储示波器

cathode ray tube：阴极射线管（电子枪显像管）

analog-to-digital converter：模拟/数字转换器

BNC connector：BNC 接头

coaxial cable：同轴电缆

attenuation：衰减

frequency response：频率响应

zoom：局部放大

time-varying magnetic fields：时变磁场

inductor：电感

sawtooth waveform：锯齿波波形

triangular waveform：三角波波形

rectangular wave：矩形波

duty cycle：占空比

peak-to-peak：峰到峰（波形峰-峰值）

trough：波谷

crest：波峰

root mean square：均方根（缩写为 RMS）

signal-to-noise ratio：信噪比（缩写为 SNR,通常以分贝为单位）

DC coupling：直流耦合

Fourier analysis：傅里叶分析

the n-th harmonic：n 次谐波

fundamental frequency：基频（相对谐波而言）

dB：分贝（deci-Bel）

〔Translations〕

1. Analog multimeters use a microammeter with a moving pointer to display readings. Digital multimeters have numeric displays and have made analog multimeters obsolete as they are cheaper, more precise, and more physically robust than analog multimeters.

   模拟万用表使用带活动指针的微电流表指示读数.数字万用表带有数字显示

器,使模拟万用表显得过时,因为数字表比模拟表更便宜、更精确,而且物理上更有耐受性.

2. The measured signal（device）will be connected into the multimeter through these terminals. Different measurement objects have different connection methods.

被测信号(设备)通过这些端口连接到万用表.不同的被测对象有不同的连接方式.

3. Early oscilloscopes used cathode ray tubes（CRTs）as their display element （hencethey were commonly referred to as CROs）and linear amplifiers for signal processing. CROs were later largely superseded by digital storage oscilloscopes （ DSOs ） with thin panel displays，fast analog-to-digital converters（ADC）and digital signal processors.

早期的示波器使用阴极射线管作为显示单元(因此它们一般又被称为 CRO),线性放大器用作信号处理.CRO 后来普遍地被带有纤薄显示器、高速模拟-数字转换器和数字信号处理器的数字存储示波器(DSO)所替代.

4. Digital oscilloscope can be adjusted so that repetitive signals can be observed as a persistent waveform on the screen. A storage oscilloscope can capture a single event and display it continuously，so the user can observe events that would otherwise appear too briefly to see directly.

数字示波器可以通过调节,使重复的信号呈现为显示屏上的稳定波形,以便观察.存储示波器可以捕获单个(波形)事件并持续显示,由此,使用者可以观察这些在其他情况下出现过于短暂而不能直接看到的事件.

5. Attach oscilloscope probes or BNC cables to these BNC connectors. In the 1000X-Series oscilloscopes, the analog channel inputs have 1 MΩ impedance，and there is no automatic probe detection，so you must properly set the probe attenuation for accurate measurement results.

将示波器探头或 BNC 电缆连接到这些 BNC 接头.在 1000X 系列的示波器上,模拟通道输入端有 1 MΩ 的电阻,且没有探头的自动检测(装置),因此你必须设置适合的探头衰减,以便获得精确的测量结果.

6. Do not connect a host computer to the oscilloscope's USB host port. A host computer sees the oscilloscope as a device. When needed，connect the host computer to the oscilloscope's device port（on the rear panel）.

不要将计算机主机连到示波器的 USB 主控端端口.主控计算机将示波器视为一个设备.如果需要,主控计算机可连接到示波器的设备端 USB(在示波器背面).

7. This cable has two main benefits：it protects the signal from external electromagnetic interference，improving accuracy for low-level signals；and

it has a lower inductance than flying leads，making the probe more accurate for high-frequency signals.

这种电缆有两个主要的好处：它保护信号免受电磁干扰，改善小信号的测量精度；它比导线直连的电感更小，对于高频信号测量更精确.

8. An improperly compensated oscilloscope probe will result in inaccurate high-frequency measurements.

补偿失当的示波器探头会导致高频测量的不精确.

9. Most oscilloscope probes can be switched between ×1 and ×10. As the ×10 probe attenuates the signal by a factor of ten，the signal entering the scope itself will be reduced. This has to be taken into account. Some oscilloscopes automatically adjust the scales according to the probe present，although not all are able to do this. It is worth checking before reading on the oscilloscope.

大多数示波器探头可以在×1 和×10 之间切换. 由于×10 挡位对信号有一个 10 倍的衰减因子，进入示波器的信号本身就已经减小了，必须考虑这一因素. 有些示波器可以根据探头挡位自动调节标度，但并非所有示波器均如此. 在示波器读数之前检查一下（探头）是值得的.

10. In addition to producing sine waves, function generators may（typically）produce other repetitive waveforms including sawtooth and triangular waveforms，square waves，and pulses.

除了产生正弦波，函数信号发生器还可（典型地）产生其他重复性的波形，包括锯齿波、三角波、方波和脉冲信号.

11. Overvoltage protection of the output channel will take effect once any of the following conditions is met.

一旦下述任一条件满足，输出通道的过压保护将生效.

12. RMS, or root mean square voltage（also called effective voltage），is a method of denoting an AC waveform as an equivalent voltage which represents the DC voltage value that will produce the same power dissipation. In other words, the RMS value allows an AC waveform to be specified as a DC, because it is the equivalent DC voltage that delivers the same amount of power to a load in a circuit as the AC signal does over its cycle. The RMS voltage of a periodic signal $v(t)$ is calculated by ...

RMS，或称均方根（也叫有效值）电压是将交流波形表示为能产生相同功耗的等效直流量的一种方法. 换句话说，RMS 值允许将交流波形（的特征）用一个直流量代替，因为该直流量在电路中向负载输送与交流量整个周期相同的功率. 交流信号 $v(t)$ 的 RMS 电压计算如下……

13. The generator outputs a rectangular wave of different duty cycles. Use the

FFT function of oscilloscope to display spectrum of the signal. Record the fundamental，second and third harmonic amplitudes L_1，L_2 and L_3 in dB scale.

信号发生器输出不同占空比的矩形波. 用示波器的 FFT 功能显示信号的频谱. 记下基频、二次谐波和三次谐波的 dB 幅度值 L_1、L_2 和 L_3.

2

Kirchhoff's Circuit Laws
基尔霍夫定律

To simplify a circuit, a single equivalent resistance can be found when two or more resistors are connected together in either series, parallel or combinations of both, and that these circuits obey Ohm's law. However, in some complex circuits such as bridge or T-networks, we can not simply use Ohm's law alone to find the voltages or currents circulating within the circuit. For these cases, Kirchhoff's circuit laws are useful tools in circuit analysis.

Kirchhoff's[1] circuit laws are two equalities that deal with the current and potential difference (commonly known as voltage) of electrical circuits. Both of Kirchhoff's laws can be understood as corollaries of Maxwell's[2] equations in the low-frequency limit. They are accurate for DC circuits, and for AC circuits at frequencies where the wavelengths of electromagnetic radiation are very large compared to the circuits.

2.1 Kirchhoff's Current Law (基尔霍夫电流定律)

Kirchhoff's current law (also known as Kirchhoff's first law) states the following:

The algebraic sum of currents in a network of conductors meeting at a node is zero.

Recalling that current is a signed quantity (positive or negative) reflecting direction towards or away from a **node** [or a junction. See Figure 2.1(a)], this

---

① Gustav Robert Kirchhoff (Mar. 12, 1824 – Oct. 17, 1887), German physicist.

② James Clerk Maxwell (Jun. 13, 1831 – Nov. 5, 1879), Scottish mathematician and physicist.

principle can be succinctly stated as

$$\sum_{k=1}^{N} I_k = 0$$

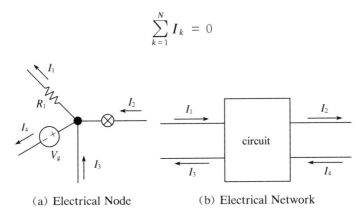

(a) Electrical Node   (b) Electrical Network

Figure 2.1 Kirchhoff's Current Law

图 2.1 基尔霍夫电流定律

where N is the total number of branches with currents flowing towards or away from the node. Kirchhoff's current law is a consequence of charge conservation.

The term **node** in an electrical circuit generally refers to a connection or junction of two or more current carrying paths or elements such as cables and components. For a closed electrical circuit, Kirchhoff's current law also applies as in Figure 2.1(b). In this case,

$$I_1 - I_2 - I_3 + I_4 = 0$$

2.2 Kirchhoff's Voltage Law (基尔霍夫电压定律)

Kirchhoff's voltage law (also known as Kirchhoff's second law) states the following:

The algebraic sum of all voltages in a loop must be equal to zero.

Similarly to Kirchhoff's current law, the voltage law can be formulated as

$$\sum_{k=1}^{N} V_k = 0$$

where N is the total number of voltages measured in a loop (see Figure 2.2). Kirchhoff's voltage law is a consequence of both charge conservation and the conservation of energy.

Kirchhoff's laws have limitations.

Kirchhoff's current law is based on the assumption that the net charge in any wire, junction or lumped component is constant. Whenever the electric field between parts of the circuit is not negligible, such as when two wires are capacitively coupled, this may not be the case. This occurs in high frequency AC circuits, where the lumped element model is no longer applicable. For example, in a transmission line, the charge density in the conductor will constantly be oscillating.

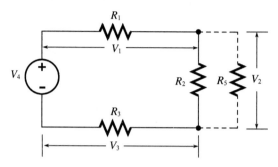

Figure 2.2　Kirchhoff's Voltage Law
图 2.2　基尔霍夫电压定律

On the other hand, the voltage law relies on the fact that the action of time-varying magnetic fields are confined to individual components, such as inductors. In reality, the induced electric field produced by an inductor is not confined, but the leaked fields are often negligible.

2.3　Example of Calculation (计算举例)

Assume an electric network consisting of two voltage sources and three resistors (see Figure 2.3).

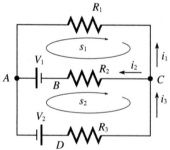

Figure 2.3　KCL Example
图 2.3　电流定律示例

For node C, according to the Kirchhoff's current law,

$$- i_1 - i_2 + i_3 = 0$$

Applying the Kirchhoff's voltage law to the *closed circuit s*1, and substituting for voltage using Ohm's law, gives

$$V_1 - R_1 i_1 + R_2 i_2 = 0$$

To the loop $s2$, gives

◀◀◀

$$+ R_3 i_3 - V_2 - V_1 + R_2 i_2 = 0$$

This yields a system of linear equations in i_1, i_2, i_3

$$\begin{cases} - i_1 - i_2 + i_3 = 0 \\ V_1 - R_1 i_1 + R_2 i_2 = 0 \\ R_3 i_3 - V_2 - V_1 + R_2 i_2 = 0 \end{cases} \tag{2.1}$$

When V and R are known, the equation can be fully solved.

2. 4 Simulation Tool—SPICE (仿真工具——SPICE)

SPICE (Simulation Program with Integrated Circuits Emphasis) is a general-purpose, open-source analog electronic circuit simulator. It was originally developed at the Electronics Research Laboratory of the University of California, Berkeley. SPICE is used to check the integrity of circuit designs and to predict circuit behavior. One of its successors, PSPICE, is the most prominent commercial version, whereas NGSPICE is one of the open-source version of SPICE. An example of using NGSPICE to analyze circuit in Figure 2.3 is given below (You need to install the NGSPICE software first).

NGSPICE requires you to describe your circuit as a netlist. A netlist is defined as a set of circuit components and their interconnections. To create a netlist file, you need first to mark all the nodes in the circuit (A, B, C, D in Figure 2.3) and take one of the node as ground ($V_A = 0$ V, for example). Listing 2.1 is the netlist file "kcldemo. cir" that describe the circuit.

Listing 2.1   Netlist for Circuit in Figure 2.3

清单 2.1   电路图 2.3 的网连表单

```
1  Kirchhoff Current Law (demo)
2
3  * Resistor R1 (1 kOhm) is connected between node C and 0 (ground)
4  R1    0    C    1k
5  R2    B    C    2k
6  R3    D    C    1k
7  V1    0    B    5V
8  V2    0    D    -3V
9  .end
```

Here is the explanation line by line to the file:

- Line 1. Title line. The title line must be the first in the circuit description file. Its contents are printed verbatim as the heading for each section of output.
- Line 3. Lines starting with " * " are comments.

- Line 4 – 8. Element connection. The first string ("R1", for example) is element name, followed by connection nodes and its attributes. The first letter of the element instance name specifies the element type. NGSPICE specifies a set of letters representing the element type: "R" for resistor, "C" for capacitor, "V" for voltage source, "I" for current source, "Q" for BJT (Bipolar Junction Transistor) transistor, "D" for diode, etc. An instance name may contain one or more characters. Hence, "R", "R1", "ROUT" are valid resistor names, "Vcc", "Vs" are valid voltage source names. Each circuit must have a ground node which is named "0". Therefore, "R1 0 C 1k" means that a resistor of 1 kΩ is connected between node C and ground (node A).

  Element names and node names are **case insensitive**.
- Line 9. End of section. The last line must be ". end".

To simulate the circuit, run NGSPICE in a terminal to load ". cir" file. The program prints some information about the software and circuit, and gives a prompt at the end:

```
$ ngspice kcldemo.cir
...
...
ngspice 1 - >
```

By issuing "listing" command, you may check the circuit description read by the simulator:

```
ngspice 1 - > listing
     kirchhoff current law ( demo)

     1 : kirchhoff current law ( demo)
     2 : . global gnd
     4 : r1    0    c    1k
     5 : r2    b    c    2k
     6 : r3    d    c    1k
     7 : v1    0    b    5
     8 : v2    0    d    - 3
    10 : . end
ngspice 2 - >
```

To run the DC simulation of the circuit, issue the following command:

```
ngspice 2  - >   op
Doing analysis at TEMP = 27.000000 and TNOM = 27.000000

No. of Data Rows : 1
ngspice 3 - >
```

Then you can print voltages and currents of the circuit：

```
ngspice 3 - > print all
b = - 5.00000 e+ 00
c = 2.000000 e- 01
d = 3.000000 e+ 00
v1 # branch = - 2.60000 e- 03
v2 # branch = 2.800000 e- 03
ngspice 4 - >
```

You can also print individual voltage or current by using "print allv", "print alli", or even print a specified node "print v(b)".

According to the results printed above, we have $V_B = -5$ V, $V_C = 0.2$ V, $V_D = 3$ V, $i_2 = 2.6$ mA(V_1 branch), $i_3 = 2.8$ mA (V_2 branch) and $i_1 = i_3 - i_2 = 0.2$ mA. Put $R_1 = 1$ kΩ, $R_2 = 2$ kΩ, $R_3 = 1$ kΩ, $V_1 = 5$ V, $V_2 = 3$ V in equation (2.1), you may check if the results are correct.

Once the netlist file is modified, you can use "source" command to reload it in NGSPICE. To quit the simulator, just issue "quit" command：

```
ngspice 4 - > source kcldemo.cir

Circuit: kirchhoff current law ( demo)

ngspice 5 - > quit
$ _
```

NGSPICE is a professional, powerful mixed-signal electronic circuit simulator and is free of charge. For more infomation about this software, please read the **Ngspice User's Manual**.

2.5 Objective (实验目的)

The purpose of this experiment is to investigate the principles of Kirchhoff's current law and Kirchhoff's voltage law.

2.6 Experimental Procedure (实验步骤)

2.6.1 Preparations (实验准备)

Power Supply

In this experiment, a DC power supply GPD-X303S is used to produce voltage source and current source. The GPD-X303S series has at least 2 independent adjustable voltage outputs. The outputs can automatically switch between constant voltage mode (CV) and constant current mode (CC), according to load condition:

- When the current level is smaller than the output setting, the device operates in constant voltage mode and the indicator on the front panel turns green (CV). The voltage level is kept at the setting and the current level fluctuates according to the load condition until it reaches the output current setting.

- When the current level reaches the output setting, the device starts operating in constant current mode and the indicator on the front panel turns red (CC). The current level is kept at the setting, but the voltage level becomes lower than the setting in order to suppress the output power level from overload. When the current level becomes lower than the setting, it goes back to the constant voltage mode and the indicator turns green (CV).

⚠ When the device is used as voltage power supply, red indicator usually means there is a short in the circuit.

Breadboard

Breadboard is a construction base used to build semi-permanent prototypes of electronic circuits (Figure 2.4).

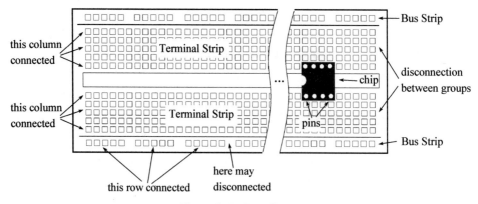

Figure 2.4　Breadboard
图 2.4　面包板

In the main area (terminal strips), holes on the breadboard are connected electrically in columns. These columns are separated from each other and disconnected. Two terminal strips are separated by a trough and are also disconnected. Upper and lower rows (bus strips) are not connected with others, but

◀◀◀

each row is connected internally for its own. Some manufacturers connect all holes in a row, others just connect groups of, for example, 15 or 20 consecutive holes in a row. Bus strips are often used to connect power supply, for example, upper row is connected to $+5$ V and lower row to ground.

2.6.2 Operations (实验操作)

1. Refer to Figure 2.5(a), build your circuit on the breadboard.

   Record your measurement parameters and write them in Table 2.1. You may either measure I_1, I_2 and I_3 using ammeter, or V_{AB}, V_{CB} and V_{BE} using voltmeter, and calculate the corresponding currents by applying Ohm's law. Note the polarity when you are reading.

Table 2.1 Kirchhoff's Current Law

表 2.1 基尔霍夫电流定律

	V_{AB}(V)	V_{CB}(V)	V_{BE}(V)	I_1(mA)	I_2(mA)	I_3(mA)
R_1						
Diode	—					

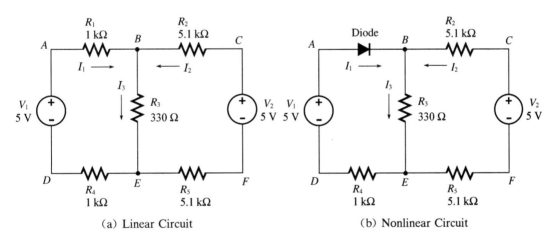

(a) Linear Circuit   (b) Nonlinear Circuit

Figure 2.5 Verification of Kirchhoff's Current Law

图 2.5 验证基尔霍夫电流定律

2. Replace resister R_1 with a diode, build your circuit according to Figure 2.5 (b), repeat the above measurement and write your record in Table 2.1. As diode is a **nonlinear element**, you can't apply Ohm's law directly to the diode since the resistance is unknown in this case. You can measure the voltage across R_4 instead.

   What conclusion do you draw from above tests?

3. Refer to Figure 2.6(a), build your test circuit on the breadboard.

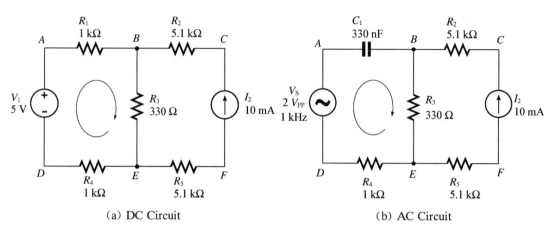

(a) DC Circuit　　　　　　　　　　　　(b) AC Circuit

Figure 2.6　Verification of Kirchhoff's Voltage Law

图 2.6　验证基尔霍夫电压定律

4. Measure the voltages required in Table 2.2 and fill them in the table.

5. Replace R_1 with an 150 nF capacitor, and V_1 with sinusoidal signal generator V_S. Apply 1 kHz/ 2 V_{PP} sine wave to the circuit. Measure the voltages and phases φ_{AB} across the capacitor using oscilloscope and fill your recorded data in Table 2.2.

Table 2.2　Kirchhoff's Voltage Law

表 2.2　基尔霍夫电压定律

Source	V_{AB} (V)	V_{BE} (V)	V_{ED} (V)	φ_{AB}
5 V DC			—	
2 V AC				

⚠ Do not apply V_S and V_1 at the same time! Voltage sources are not allowed to be connected in parallel.

Check your measurement. Does the AC circuit obey Kirchhoff's law?

The netlist file of Figure 2.6(b) is given below. NGSPICE loads the script file, runs ".control" section and plots waveforms at node A and B (Figure 2.7). You may find the phase difference across the capacitor.

Listing 2.2   Netlist of Figure 2.6(b)

清单 2.2   电路图 2.6(b)的网连表单

```
1   Kirchhoff Voltage Law, AC
2
3   .tran  10u  5m       ; total 5ms, with timestep of 10us
4
5   VS    A     0    ac  1  sin(0 1 1k)
6
7   C1    A     B    150n
8   R2    B     C    5.1k
9   R3    B     E    330
10  R4    0     E    1k
11  R5    E     F    5.1k
12  I2    F     C    10 mA
13  .control
14  set color0= white    ; background color
15  set color1= black    ; grid and text color
16  run
17  plot v(A)   v(B)
18  .endc
19  .end
```

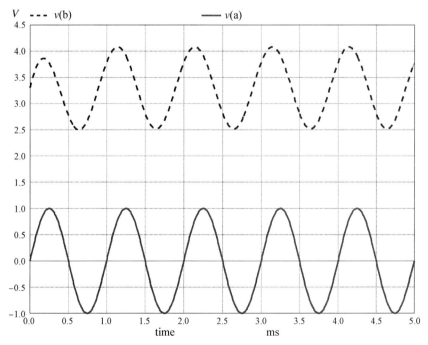

Figure 2.7   Waveform at Node *A* and *B* of Figure 2.6(b)

图 2.7   图 2.6(b) *A*、*B* 两点处的电压波形

6. Remove voltage source and short circuit terminals A and D, write down the voltage V_{BE}; Add the voltage source back and disconnect the current source, write down the voltage V_{BE} again [Figure 2.8 (a) and (b)].

Add these two values of measurement. Compare the result with that of in Table 2.2, what can you find? Are there any theorem to explain this phenomenon?

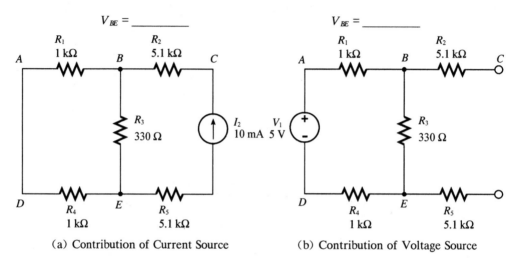

(a) Contribution of Current Source   (b) Contribution of Voltage Source

Figure 2.8 Experiment Circuit

图 2.8 实验电路

［实验简介］

基尔霍夫定律是分析低频电路的重要工具. 本实验学习使用该理论分析电路的方法，了解该理论的适用范围.

［Vocabulary］

corollary：推论

lumped element：集总元件

node：结点

junction：结点

succinctly：简洁地

negligible：可忽略的

capacitively coupled：通过电容耦合

time-varying magnetic fields：时变磁场

inductor：电感

closed circuit：闭合电路

netlist：网连表单

polarity：极性

nonlinear：非线性的

bipolar junction transistor：双极结型晶体管

case insensitive：字母大小写无关

breadboard：多孔板，面包板

overload：过载

　　集总元件(lumped element)是指元件大小远小于电路工作频率所对应的电磁波波长时，对所有元件之统称．这种情况下，对于信号而言，任何时刻元件特性始终保持固定，与频率无关．相反地，若元件尺寸与电路工作频率所对应的波长相当或者更大，则当信号通过元件之时，元件本身各点的特性将因信号的变化而有所不同，此时不能将元件整体视为均匀特性的单一体，而应称为分布元件(distributed element)．例如在研究微波电路时，传统的纯电阻导线很可能会成为具有电感及电容特性的复杂组合体．

〖Translations〗

1. Both of Kirchhoff's laws can be understood as corollaries of Maxwell's equations in the low-frequency limit.

   基尔霍夫的两个定律都可被理解为麦克斯韦方程在低频极限的推论．

2. The algebraic sum of currents in a network of conductors meeting at a node is zero.

   在导体构成的网络汇聚点上，电流的代数和为零．

3. Recalling that current is a signed quantity（positive or negative）reflecting direction towards or away from a node［see Figure 2.1(a)］, this principle can be succinctly stated as ...

   不要忘了，电流是有符号的量（正或负），（符号）反映电流流入或者流出结点［见图2.1(a)］，这一原理可以简洁地表述为……

4. The current law is based on the assumption that the net charge in any wire, junction or lumped component is constant. Whenever the electric field between parts of the circuit is not negligible，such as when two wires are capacitively coupled，this may not be the case.

   电流定律基于这样的假设：导线、结点以及集总元件中的净电荷是恒定的．一旦部分电路之间的电场不能忽略，如两根导线之间发生了电容耦合（效应），就不是这样的情况了（不满足假设条件）．

5. NGSPICE requires you to describe your circuit as a netlist. A netlist is defined as a set of circuit components and their interconnections.

   NGSPICE 要求你将电路用一个网络连接清单描述．连接清单定义了一组电路元件以及它们之间的连接关系．

6. When the current level reaches the output setting, the device starts operating

in constant current mode and the indicator on the front panel turns red (CC). The current level is kept at the setting, but the voltage level becomes lower than the setting in order to suppress the output power level from overload.

当电流达到输出设置时,仪器开始处于恒流源模式,前面板上的指示灯变成红色(CC).(在此工作方式下)电流水平保持设定值,但是电压值低于设定值,以降低输出功率水平,避免过载.

3

Equivalent Circuits
等效电路

Thévenin's[1] theorem and Norton's[2] theorem are widely used for circuit analysis simplification and to study circuit's **initial-condition** and **steady-state response**.

3.1 Thévenin's Theorem (戴维宁定理)

In terms of direct current resistive circuits (see Figure 3.1), Thévenin's theorem states as following:

For any linear electrical network containing only voltage sources, current sources and resistances can be replaced at terminals A and B by an equivalent combination of a voltage source V_{th} in a series connection with a resistance R_{th}.

The main points of analyzing a circuit using Norton's theorem:

1. The equivalent voltage V_{th} is the voltage obtained at terminals A and B of the network with terminals A and B open circuited.

2. The equivalent resistance R_{th} is the resistance that the circuit between terminals A and B would have if all ideal voltage sources in the circuit were replaced by a short circuit and all ideal current sources were replaced by an open circuit.

3. If terminals A and B are short-circuited, the current flowing from A to B will be V_{th}/R_{th}. This means that R_{th} could alternatively be calculated as V_{th}

---

① Léon Charles Thévenin (Mar. 30, 1857 – Sept. 21, 1926), French telegraph engineer.

② Edward Lawry Norton (Jul. 28, 1898 – Jan. 28, 1983), American engineer and scientist worked at Bell Labs. He received a B. S. degree from MIT in 1922 and an M. A. degree from Columbia University in 1925.

divided by the short-circuit current between *A* and *B* when they are connected together.

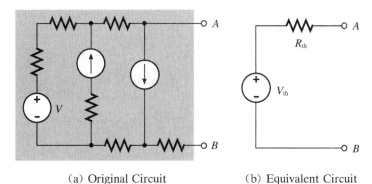

(a) Original Circuit (b) Equivalent Circuit

Figure 3.1 Thévenin Equivalent Circuit

图 3.1 戴维宁等效电路

3.2 Norton's Theorem (诺顿定理)

In direct-current circuit theory, Norton's theorem is a simplification that can be applied to networks made of linear time-invariant resistances, voltage sources, and current sources (see Figure 3.2).

At a pair of terminals of the network, it can be replaced by a current source and a single resistor in parallel.

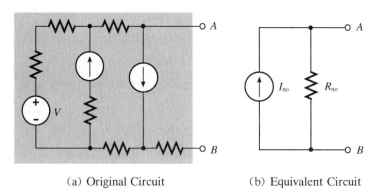

(a) Original Circuit (b) Equivalent Circuit

Figure 3.2 Norton Equivalent Circuit

图 3.2 诺顿等效电路

1. The equivalent current I_{no} is calculated as the current flowing at the terminals into a short circuit (zero resistance between *A* and *B*).

2. The Norton resistance R_{no} is found by calculating the output voltage produced with no resistance connected at the terminals; equivalently, this is the resistance between the terminals with all independent voltage sources

short-circuited and independent current sources open-circuited.

In circuit theory terms, both theorems allow any one-port network to be reduced to a single voltage source or current source and a single impedance. The theorems also apply to frequency domain AC circuits consisting of reactive and resistive impedances. It means the theorems apply for AC in an exactly same way to DC except that resistances are generalized to impedances.

A Norton equivalent circuit is related to the Thévenin equivalent by the following equations

$$\begin{cases} R_{th} = R_{no} \\ V_{th} = I_{no} R_{no} \\ \dfrac{V_{th}}{R_{th}} = I_{no} \end{cases} \qquad (3.1)$$

3.3 Example of Calculation (计算举例)

Look at Figure 3.3(a), for example. Take the point B as ground ($V_B = 0$). When terminals A and B are open-circuited, the voltage at node C is

$$V_C = \frac{R_3 + R_4}{R_1 + (R_3 + R_4)} V_S = 7.5 \text{ V}$$

Short-circuit V_S, the resistance between A and B is

$$R_{AB} = R_{th} = R_1 \parallel (R_3 + R_4) + R_2 = 2 \text{ k}\Omega$$

Thus, we get Thévenin equivalent circuit of Figure 3.3(b).

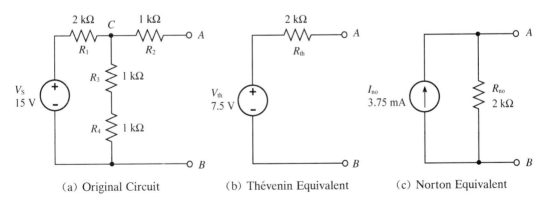

(a) Original Circuit　　(b) Thévenin Equivalent　　(c) Norton Equivalent

Figure 3.3　Equivalent Circuits

图 3.3　等效电路

To find the Norton equivalent, short-circuit terminals A and B in Figure 3.3 (a). The current flowing through R_1 is

$$I_1 = \frac{V_s}{R_1 + (R_3 + R_4) \parallel R_2} = 5.625 \text{ mA}$$

the current flowing through R_2 is

$$I_2 = \frac{1/R_2}{1/(R_3 + R_4) + 1/R_2} I_1 = 3.75 \text{ mA}$$

According to equation (3.1), we have $R_{no} = 2 \text{ k}\Omega$. The result is shown in Figure 3.3 (c).

3.4   Superposition Theorem (叠加定理)

The superposition theorem states:

For a linear system (notably including the subcategory of time-invariant linear systems), the response (voltage or current) in any branch of a bilateral linear circuit having more than one independent source equals the algebraic sum of the responses caused by each independent source acting alone, where all the other independent sources are replaced by their internal impedances.

To analyze the contribution of each individual source in a circuit with multiple sources, all of the other sources must be "turned off" (set to zero) first by:

- Replacing all other independent voltage sources with a short circuit, as an ideal voltage source has zero internal impedance.
- Replacing all other independent current sources with an open circuit, as the internal impedance of an ideal current source is infinite.

This procedure is followed for each source in turn, then the resultant responses are added to determine the true operation of the circuit.

The theorem is applicable to linear networks (time varying or time invariant) consisting of independent sources, linear dependent sources, linear passive elements (resistors, inductors, capacitors) and linear transformers.

The superposition theorem is very important in circuit analysis. It is used in converting any circuit into its Norton equivalent or Thévenin equivalent.

3.5   Objective (实验目的)

The purpose of this experiment is to investigate Thévenin and Norton equivalent circuit. Use the principle of superposition along with Thévenin's and Norton's theorems to reduce complex circuit to simple voltage or current source model.

本实验帮助理解等效电路原理,掌握等效电路的分析方法.

3. 6   Experimental Procedure (实验步骤)

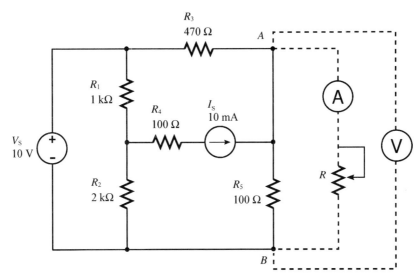

Figure 3.4   Verification of Thévenin's Theorem

图 3.4   验证戴维宁定理

1. Refer to Figure 3.4 to build your circuit on the breadboard. Connect a potentiometer R across terminals A and B as a variable load (dashed part). Sketch the circuit in your lab notebook. Since you will be doing a lot of circuit variations，it is important to get into the habit of always sketching the circuit in your lab notebook. Always include the supply voltages when sketching circuit.

2. Adjust the variable resistor R，and measure the voltage V between the terminals A and B and the current I flowing through R with different resistance values. The value of current I can also be calculated by measuring resistance R and using Ohm's law. Fill your records in Table 3.1.

Table 3.1   Verification of Thévenin's Theorem

表 3.1   验证戴维宁定理

V (V)	1.0	1.2	1.4	1.6	1.8	2.0	2.2	2.4
R (kΩ)								
I (mA)								
V' (V)	1.0	1.2	1.4	1.6	1.8	2.0	2.2	2.4
R (kΩ)								
I' (mA)								

3. Sketch Thévenin's equivalent circuit as in Figure 3.1(b), calculate V_{th} and Thévenin resistance R_{th}.

$V_{th} =$ _____ V, $R_{th} =$ _____ Ω

4. Build the Thévenin's equivalent circuit as Figure 3.1(b) based on your calculation above. Connect a potentiometer R between terminals A and B as a load, and measure its voltage V' and current I' according to the different resistance values if ammeter is not available. Fill the recorded data in Table 3.1.

5. Based on your measurements, plot the current vs. voltage curve $V = f(I)$ and $V' = f(I')$ of the two circuits.

6. Try to write a netlist script file for the circuit (Figure 3.4), use simulation tool to verify your results.

[实验简介]

　　等效电路是指可以表示某个给定电路所有特性的理论电路. 在电路分析中,常常希望将一个复杂电路用相对简单的形式表示,以简化计算. 这是等效电路最常见的应用场景. 对于更复杂的情况,通过合理近似形成的等效电路,在后续的三极管电路和运算放大器电路中也是一种重要的分析方法.

　　等效电路仅在端口提供与原电路相同的特性,不考虑电路内部单元和结点. 多数情况下,等效电路只存在于理论分析中,是理论上的"等效",不是真实电路元件的"替换". 本实验的目的是帮助读者理解等效电路的原理,掌握通过等效电路方法分析复杂电路的手段,并不是在分析电路时真的发生元件替换.

[Vocabulary]

equivalent circuit:等效电路

resistive:电阻性的

reactive:电抗性的

steady-state response:稳态响应

short circuit:短路

open circuit:开路

direct current circuit:直流电路（常缩写为 DC）

impedance:阻抗

superposition theorem:叠加定理

time-invariant system:非时变系统,时不变系统

alternating current circuit:交流电路（常缩写为 AC）

breadboard:多孔板,面包板

potentiometer:电位器,可变电阻

〔**Translations**〕

1. It means the theorems apply for AC in an exactly same way to DC except that resistances are generalized to impedances.

   这意味着该理论对于交流电路同样适用,只要把电阻推广到阻抗.

2. The Norton resistance R_{no} is found by calculating the output voltage produced with no resistance connected at the terminals; equivalently, this is the resistance between the terminals with all independent voltage sources short-circuited and independent current sources open-circuited.

   诺顿电阻 R_{no} 通过两端不接电阻时的输出电压计算得到.等效地,这是两端之间所有独立电压源短路、独立电流源开路的电阻.

3. To analyze the contribution of each individual source in a circuit with multiple sources, all of the other sources must be "turned off"(set to zero) first by ...

   为了分析带有多个电源的电路中的每个独立电源的贡献,先通过下面的方法"关闭"(设置为 0)所有其他电源……

4. Sketch the circuit in your lab notebook. Since you will be doing a lot of circuit variations, it is important to get into the habit of always sketching the circuit in your lab notebook. Always include the supply voltages when sketching circuit.

   把电路图画在你的实验记录本上.你将做大量不同的电路实验,因而养成总是把电路图记在你的实验记录本上的习惯是十分重要的.画电路时不要忘记电源电压.

5. Replacing all other independent voltage sources with a short circuit, as an ideal voltage source has zero internal impedance.

   将其他所有独立电压源用短路线替代,因为理想电压源的内阻为零.

Transistor Amplifiers
晶体管放大器

Amplifiers are widespread in all areas of electronic engineering. Understanding how transistors function is important to anyone interested in understanding modern electronics. During this experiment, you will learn some key operating features of BJT amplifier circuits. These circuits will provide information to help you in further studies of these important topics.

4.1 Basic Amplifier Configurations (放大器基本构成形式)

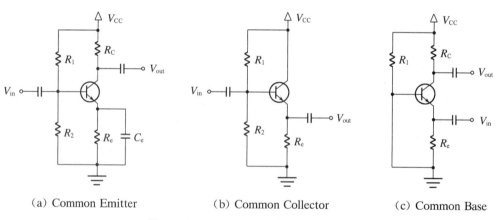

(a) Common Emitter   (b) Common Collector   (c) Common Base

Figure 4.1 Three Typical Configurations
图4.1 三种典型的放大器配置

A BJT (Bipolar Junction Transistor) has three terminals: base, collector, and emitter. Depending on which terminal is "common", three basic configurations are available: common-emitter, common-collector and common-base. The word "common" here means "AC-grounded". Each of these configurations has widely different characteristics and is used for different purposes. The main characteristics of the three basic configurations are listed in Table 4.1.

Table 4.1   Characteristics of Amplifier in Different Configurations

表 4.1   不同配置放大器的特性

	Common Emitter	Common Collector	Common Base
Voltage Gain	$\sim10^2$	<1	$\sim10^2$
Current Gain	$\sim10^2$	$\sim10^2$	<1
Phase	inverted	not inverted	not inverted
Input Impedance	$\sim$kΩ	~100 kΩ	~10 Ω
Output Impedance	$\sim$kΩ	~100 Ω	$\sim$MΩ

Take a look at the common-emitter amplifier [Figure 4.1(a)], for example. At first glance, none of the terminals is grounded. Recall however that a capacitor is a short circuit for AC signal. Thus, here the emitter is AC-grounded, and is then "common". By this configuration, the input signal is applied at the base, and the output signal is sampled at the collector.

You are encouraged to analyze Figure 4.1(b) and Figure 4.1(c) using same methods. Note that a constant voltage source is seen as ground for AC signals.

4.2   DC Performance（直流性质）

We will start our study of BJT amplifiers with the **common emitter amplifier**, commonly abbreviated as the CE amplifier, as it is the most commonly used BJT amplifier. Understanding the CE amplifier already gives you a powerful reference when analyzing circuits.

Our goal here is to understand what each component in this classical circuit is used for, and what are the characteristics of the amplifier. More specifically, we will derive voltage gain, current gain, input resistance, and output resistance.

4.2.1   AC Coupling（交流耦合）

Let's take a closer look at the schematic (Figure 4.2), as there are plenty of interesting things going on here.

First, notice that the input signal v_{in} is not applied directly to the base. Instead, it passes through a capacitor C_1 first. Recall that a capacitor acts like a short circuit for AC signals, but like an open circuit for DC, this allows the input AC signal to pass through the capacitor while stopping DC voltage from the power supply from leaking out. This is called **AC coupling** the input.

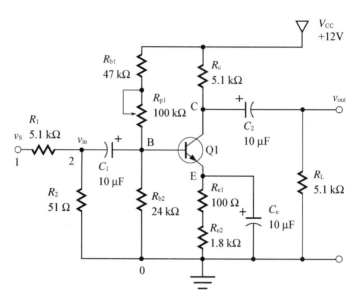

Figure 4.2   Single-Stage Amplifier

图 4.2   单级放大电路

Just like C_1, C_2 allows AC signals to pass through to the load while blocking DC signals. The output here is also AC-coupled. We seldom want DC offset at the output, but if we do, we can simply remove C_2.

By the way, the input circuit with R_1 and R_2 acts as a voltage divider, gives about 1/100 of input signal v_S to the amplifier (as long as the amplifier's input resistance $r_i \gg R_2$). As the voltage gain of the amplifier can be very large, and the output signal must be limited within V_{CC} to prevent from distortion, the input signal can't be too large. This voltage divider allows the function generator to use hundreds of millivolts of signal v_S input to the amplifier. This improves the measurement of voltage gain, as the **signal-to-noise ratio** of v_S is not too low, and the voltage gain of the amplifier is calculated simply as

$$A_v = \frac{v_{out}}{v_{in}} \approx \frac{v_{out}}{v_S} \times 100\%$$

When the amplifier is in practical use, the voltage divider is not necessary.

Some important notation should be mentioned here. Voltage at a transistor terminal is indicated by a single subscript. For instance, V_C is the collector voltage. Voltage between two terminals is indicated by a two-letter subscript: V_{BE}, for example, denotes the base-to-emitter voltage. If the same letter is repeated in subscript, that means a power-supply voltage. Therefore, V_{CC} is the (positive) power-supply voltage associated with the collector, and V_{EE} is the (negative) power-supply voltage associated with the emitter (if the emitter is not grounded).

4.2.2 Biasing the Transistor (晶体管偏置)

We know that a transistor requires a power supply to work. Here, this power supply is modeled with V_{CC}. But what are R_{b1} (associated with a potentiometer R_{p1}) and R_{b2} used for? Wouldn't the circuit work without them?

Knowing that the base-emitter junction of NPN transistors needs to be forward biased for the transistor to conduct, if we want the transistor to amplify signals all the time, the BE junction needs to be forward biased all the time. This means that we must have $V_{BE} > 0.7$ V (for silicon transistor) all the time. If we only apply v_i at the base without anything else, then we would have $v_i = V_B$. But v_i is a pure AC signal with no DC offset (the capacitor C_1 removes any DC offset if present), it will swing between two values, one positive and one negative. This will cause the transistor to turn off when $V_{BE} = v_i < 0.7$ V. The output signal will be heavily distorted.

However, if we add some DC offset to the base voltage, such that even with the swing of our input voltage $V_{BE} > 0.7$ V all the time, then we can guarantee that our transistor will be always on. This is called biasing the transistor.

In this circuit, biasing is done with a voltage divider consists of R_{b1}, R_{p1} (we will use $R'_{b1} = R_{b1} + R_{p1}$ later for simplicity) and R_{b2}. This gives the transistor a bias voltage approximately equal to

$$V_B \approx \frac{R_{b2}}{(R'_{b1} + R_{b2})} V_{CC} \tag{4.1}$$

Note that the voltage divider formula is only valid if no current is lost between the two resistances. In fact, a small amount of current I_B is sent through the base-emitter junction,

$$I_B = \frac{V_{CC} - V_B}{R'_{b1}} - \frac{V_B}{R_{b2}} \tag{4.2}$$

However, since the base current is very small (tens of μA, much less than I_B, which is several mA), formula (4.1) still gives a reasonably accurate value that we can exploit. You will be measuring this current in your experiment.

The exact voltage chosen for biasing depends on the transistor's characteristics. We will discuss it later.

4.2.3 Determination of the Quiescent Point (确定静态工作点)

With biasing, even without any input signal the transistor still conducts. There is voltage at the base, and there is current flowing through the transistor. This point of operation is called the **quiescent point**, or Q-point. The collector current at this point is called the quiescent current I_{CQ}. There is a clear way to graphically exploit this point. Let's take a look at our transistor's collector characteristic curves, which gives us the relationship $I_C = f(V_{CE})$. A simplified version of these curves is found in Figure 4.3.

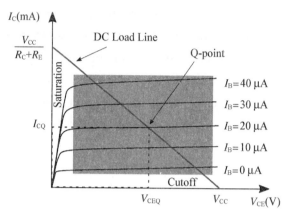

Figure 4.3   Q-point and DC Load Line

图 4.3   静态工作点和直流负载线

Let's place our quiescent point on these curves. Going back to our schematic, a KVL (Kirchhoff's Voltage Law) loop from V_{CC} to ground gives us

$$V_{CC} - R_C \cdot I_C - V_{CE} - R_E \cdot I_E = 0 \qquad (4.3)$$

C_e is a bypass capacitor that provides a low impedance path for AC current from emitter to ground, thereby removing R_E (required for good Q-point stability) from the circuit only when AC signals are considered.

We can approximate $I_C \approx I_E$. In fact, since $I_E = I_C + I_B$ (Kirchhoff's Current Law). Remember that $I_C = \beta \cdot I_B$,[①] with common values of β being in the range of 100 – 300, we have $I_B \ll I_C$. This approximation is accurate enough.

With this approximation, equation (4.3) becomes

$$V_{CC} - (R_C + R_E) \cdot I_C = V_{CE} \qquad (4.4)$$

This gives us I_C:

$$
\begin{aligned}
I_C &= \frac{V_{CC} - V_{CE}}{R_C + R_E} \\
&= \frac{V_{CC}}{R_C + R_E} - \frac{V_{CE}}{R_C + R_E}
\end{aligned}
\qquad (4.5)
$$

Equation (4.4) gives us a linear equation linking I_C and V_{CE}. We can plot this straight line on the above transistor characteristic curves. Two points are enough for

---

①   In h-parameter transistor model and in many datasheets, β is often written as h_{FE}.

us to plot it:

1. $I_C = 0$ gives us $V_{CC} = V_{CE}$.

2. When the transistor is fully on, $V_{CE} = 0$ gives us $I_C = \dfrac{V_{CC}}{R_C + R_E}$.

The quiescent point of this amplifier is somewhere along this slope. We now need to find out which curve we need to use. To select a suitable curve, we need to calculate the base current at the quiescent point I_{BQ}.

When the transistor is on, we have $V_E = V_B - V_{BE} = V_B - 0.7$ V and $I_E = V_E/R_E$. Here $R_E = R_{e1} + R_{e2}$ in Figure 4.2. Since $I_C \approx I_E$ and $I_C = \beta \cdot I_B$, we have:

$$I_B = \frac{V_B - 0.7 \text{ V}}{\beta \cdot R_E} \tag{4.6}$$

Now that we have I_B, we can select the correct curve in Figure 4.3. The intersection of the slope and the correct characteristic curve gives us the Q-point.

By choosing R_C and R_E, we limit the possibilities of the amplifier. Its operation is limited to the load line mentioned above. When input signal is absent, $I_C = I_{CQ}$ and $V_{CE} = V_{CEQ}$. This slope is called the DC **load line** of the amplifier, and depends only on R_E and R_C.

You can also find the Q-point by calculation. A KVL loop from V_{CC} to ground gives

$$V_{CC} - R_C \cdot I_C - V_{CE} - R_E \cdot I_E = 0$$

thus

$$V_{CEQ} \approx V_{CC} - (R_C + R_E) \cdot I_{CQ}$$

4.2.4 Summary to DC Performance (直流特性小结)

AC coupling through capacitors C_1 and C_2 is used to inject AC input signal and extract output signal without disturbing Q-Point.

To summarize the steps finding the Q-Point of an amplifier:

1. Find the quiescent base voltage V_{BQ} with the voltage divider formula.

2. Find the emitter voltage $V_{EQ} = V_{BQ} - 0.7$ V.

3. Find the quiescent collector current $I_{CQ} = I_{EQ} = \dfrac{V_{EQ}}{R_{EQ}}$.

4. Derive the quiescent base current $I_{BQ} = \dfrac{I_{CQ}}{\beta}$.

5. Find the Q-Point (V_{CEQ}, I_{CQ}) either graphically, with the DC load line and I_{BQ}, or numerically with the KVL loop.

4.3　AC Performance（交流特性）

So far we have only studied the DC performance of our amplifier. However, AC signals behave very differently from DC: the way they "see" different circuit elements changes. Furthermore, AC performance of an amplifier is of particular interest to us, as the signals we will be applying to our amplifier will be AC signals. To perform an AC analysis, or **small-signal analysis**, we need to:

1. ground all voltage sources.
2. open circuit all current sources.
3. replace all capacitors with a short.
4. replace all inductors with an open circuit.

If we apply these transformations to our current circuit, we may redraw the AC-circuit as in Figure 4.4 (hybrid-pi model). This is so called the equivalent circuit of **the small signal model** of the transistor.

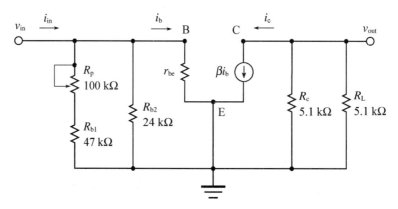

Figure 4.4　A Small Signal Model of the Transistor Amplifier

图 4.4　晶体管放大器的小信号模型

As you can see in Figure 4.4, there are three terminals corresponding to the BJT's base, collector, and emitter. The current flowing into the base is determined by the base-to-emitter voltage (v_{BE}) and r_{be}, and the collector current βi_b is generated by a current-controlled current source. The base current flows into the base, the collector current flows into the collector, and the emitter current flows out of the emitter and is the sum of the base current and the collector current (Kirchhoff's current law).

The AC resistance of BE junction r_{be} is given by

$$r_{be} = r_{bb'} + \frac{V_T}{I_{BQ}} = 200\ \Omega + \frac{25\ \text{mV}}{I_{BQ}} \tag{4.7}$$

where thermal voltage V_T is a temperature-related variable. At approximately room temperature (22 ℃ or 295 K)，V_T is about 25 mV.

4.3.1　Current Gain of the CE Amplifier(共射极放大器电流增益)

Calculating the current gain of the CE Amplifier is simple. The ratio of output current i_c, and input base current i_b, is simply the transistor's current gain β.

4.3.2　Voltage Gain (电压增益)

Look at Figure 4.4，we can now find the relationship between v_{in} and v_{out}:

$$v_{in} = r_{be} \cdot i_b$$

$$v_{out} = - R_c \parallel R_L \cdot i_c$$

The minus sign " $-$ " indicates that the output voltage is inverted from the input.

Now we can derive the voltage gain：

$$A_v = \frac{v_{out}}{v_{in}} = - \beta \frac{R_c \parallel R_L}{r_{be}} \tag{4.8}$$

Increasing the load R_L changes the flow of current，and thus changes the real gain of the amplifier. For values $R_L \gg R_c$ (open-output)，the equation (4.8) becomes $A_v \approx - \beta \dfrac{R_c}{r_{be}}$.

4.3.3　Simulation (仿真)

A simplified NGSPICE script file for Figure 4.2 is given in Listing 4.1. In this simulation，BF (β) is set to 180 for the BJT transistor model and most parameters are set default or discarded. As a result，Figure 4.5 shows the input and output waveform.

Listing 4.1　BJT Amplifier Netlist, bjtamp. cir

清单 4.1　BJT 放大器网连表单文件 bjtamp. cir

```
1   Transistor Amplifier
2
3   * Transient analysis for 5ms, with step size of 10us
4   .tran     10u     5m
5
6   Vs    1     0     ac 1 sin(0 100mV 1k)
7
8   .model 2N2222 NPN (Is= 3.0e- 14 BF= 180 Cjc= 9.0 pF Cje= 27 pF)
9
10  R1    1     2     5.1k
11  R2    2     0     51
```

```
12  C1    2     B     10u
13  Rb2   B     0     24k
14  Rb1   B     VCC   82k       ; variable
15  Q1    C     B     E     2N2222
16  Re    E     0     1.9k
17  Ce    E     0     10u
18  Rc    C     VCC   5.1k
19  C2    C     out   10u
20  RL    out   0     5.1k
21  Vcc   VCC   0     12V
22
23  .control
24  set color0= white
25  set color1= black
26  run
27  plot v(out) v(2)
28  .endc
29  .end
```

4.3.4 Input Impedance(输入阻抗)

Input impedance of a circuit is the resistance seen by the input signal when going through the circuit. **Output impedance** is the impedance seen when looking into the output port of the circuit. Input and output impedance are fundamental concepts in the field of electronics.

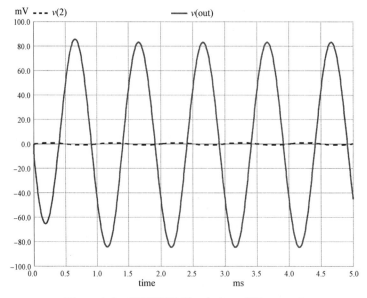

Figure 4.5   NGSPICE Simulation of Figure 4.2

图 4.5   图 4.2 的 NGSPICE 仿真波形

An amplifier, or any circuit, can be simplified as a 2, 3, or 4 port-network with a certain input impedance and a certain output impedance. Any signal flowing into an amplifier will face a certain resistance characteristic of this circuit (r_i in Figure 4.6). This is the circuit's input impedance. For frequencies in the low to mid-range, the input impedance of a BJT transistor amplifier is purely resistive.

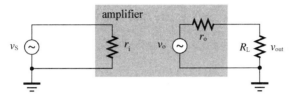

Figure 4.6　Input and Output Resistance of a 2-Port Network

图4.6　双端网络的输入和输出阻抗

A high input resistance means the circuit will draw less current from the previous stage. The higher the input resistance of a circuit is, the less impact it has on the preceding circuit. The reverse is also true. A low input resistance will draw more current from the previous stage. The lower the input resistance of a circuit is, the more impact it will have on the preceding circuit. Figure 4.7 shows an experimental method of finding input resistance of a circuit, where R_1 is a test resistor. By measuring v_S and v_i, the input resistance r_i is given by

$$r_i = \frac{v_{in}}{i_{in}} = \frac{v_{in}}{(v_S - v_{in})/R_1} = \frac{R_1}{\dfrac{v_S}{v_{in}} - 1} \tag{4.9}$$

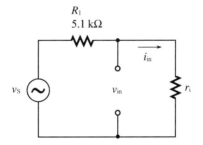

Figure 4.7　Measuring the Input Resistance

图4.7　测量输入阻抗

To find the input resistance of a circuit, we apply a test voltage v_{in} at the input, measure the current i_{in}, and calculate the ratio v_{in}/i_{in}. This is done on the small signal model of the circuit (Figure 4.4)

$$i_{in} = \frac{v_{in}}{R'_{b1} \parallel R_{b2} \parallel r_{be}}$$

which gives us

$$r_i = R'_{b1} \parallel R_{b2} \parallel r_{be} \qquad (4.10)$$

In general, for a CE amplifier, this is considered to be high, commonly in the thousands of Ohms.

4.3.5　Output Impedance (输出阻抗)

Just as input impedance is impedance seen by a signal entering the circuit's input, output impedance is seen by its load. The output impedance of a BJT transistor amplifier is also purely resistive in the low to middle frequency range.

With a low output resistance, a lot of current can be drawn from the circuit's output without changing the output voltage too much. On the other hand, a high output resistance will cause a circuit's output voltage to drop quickly when current is drawn.

To find out the output resistance of the amplifier, just remove the load R_L and apply the Thévenin's theorem to the output loop of the amplifier, i. e, replace the current source βi_b by an open circuit. The Thévenin resistance is $R_{th} = R_C$, thus the output resistance of the amplifier $r_o = R_{th} = R_C$.

The experimental way of measuring output resistance of a circuit is shown in Figure 4.8. The output resistance is given by

$$r_o = \left(\frac{v_{out}}{v_o} - 1 \right) \cdot R_L \qquad (4.11)$$

where R_L is a known load as a test resistor.

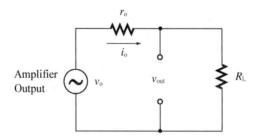

Figure 4.8　Measuring the Output Resistance
图 4.8　测量输出阻抗

4.3.6　Conclusion (结论)

Till now, we have collected sufficient information to summarize all the AC characteristics of the CE amplifier that we've derived:

- Current gain $A_i = \beta$

- Voltage gain $A_v = -\beta \dfrac{R_C \parallel R_L}{r_{be}}$

- Input resistance $r_i = R'_{b1} \parallel R_{b2} \parallel r_{be}$
- Output resistance $r_o = R_C$

⚠ Do not use ohmmeter to measure input and output resistance of an amplifier, since the ohmmeter operates in the DC mode.

4.3.7 Frequency Response (频率响应)

An electronic system has its frequency properties. Frequency response $H(j\omega)$ is a function that relates the output response to a sinusoidal input at frequency ω. In general, frequency response is complex, with real and imaginary parts. We can separate it into magnitude (called **amplitude response**) and phase component (called **phase response**). The steady-state sinusoidal frequency response of a circuit is described by the phasor transfer function. A **bode plot** is a graph of the magnitude (in dB) or phase of the transfer function with respect to frequency as shown in Figure 4.9.

Most amplifiers have relatively constant gain across a range or band of frequencies, this band of frequencies is referred to as the bandwidth (BW) of the circuit. In most electronic circuits, the word "bandwidth" refers to the range of frequencies over which the system produces a specified level of performance. In the case of frequency response, bandwidth in Hz refers to the range of frequencies more than 3 dB below the maximum amplitude gain, or 3 dB bandwidth. The -3 dB point at low frequency is called low cutoff frequency. At the high frequency side, the -3 dB point corresponds to the high cutoff frequency.

Based on the script file in Listing 4.1, you may use following commands to plot the frequency response of the amplifier. The commands set the AC analysis from 50 Hz to 50 MHz, 10 points per decade, and plot the amplitude response in dB. The result is shown in Figure 4.9. Since the BJT parameters are not set exactly as a real transistor, the result may differ from your measurement.

```
ngspice 1 - > source bjtamp.cir
ngspice 2 - > ac dec 10 50 50Meg
ngspice 3 - > plot db(v(out))
```

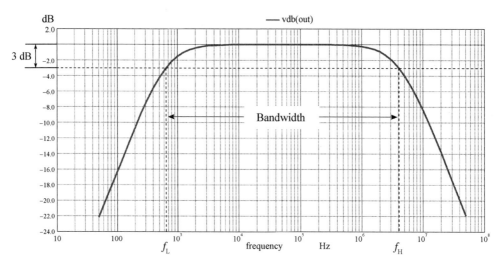

Figure 4.9   Simulation Result of Amplifier Frequency Response

图 4.9   放大器频率响应的仿真结果

4.4   Amplifier with Negative Feedback (负反馈放大器)

Feedback is the process by which a fraction of the output signal, either a voltage or a current, is used as an input. If the feedback fraction is opposite in value or phase ("anti-phase") to the input signal, then the feedback is said to be **negative feedback**. Negative feedback opposes or subtracts from the input signals giving it many advantages in the design (such as precisely gain control, reducing signal distortion, extending the bandwidth, etc.) and stabilization of control systems.

4.4.1   Closed-Loop Gain of Negative Feedback (负反馈闭环增益)

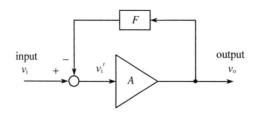

Figure 4.10   Negative Feedback Block Diagram Model

图 4.10   负反馈框图

An idealized negative-feedback amplifier consists of three elements (see Figure 4.10):

1. an amplifier with open-loop gain A.

2. a feedback network with feedback factor F.

3. a summing circuit that acts as a subtractor which combines the input and the transformed output.

Without feedback, the input voltage v'_i is applied directly to the amplifier input. The according output voltage is $v_o = Av'_i$.

Suppose the feedback loop applies a fraction F of the output v_o to one of the subtractor inputs so that it subtracts from the circuit input v_i applied to the other subtractor input. The result of subtraction applied to the amplifier input is $v'_i = v_i - F \cdot v_o$. This gives

$$v_o = A(v_i - Fv_o)$$

Then the gain of the amplifier with feedback, called the closed-loop gain, A_F is given by

$$A_F = \frac{v_o}{v_i} = \frac{A}{1 + AF} \tag{4.12}$$

For circuit with deep negative feedback, $AF \gg 1$, then $A_F \approx 1/F$. This means the closed-loop gain of the circuit is simply determined by the feedback network.

4.4.2 Feedback Topologies (反馈类型)

The feedback topology depends on whether we amplify voltage or current, and on whether we are sampling voltage or current at the output for feedback use. There are four different configurations.

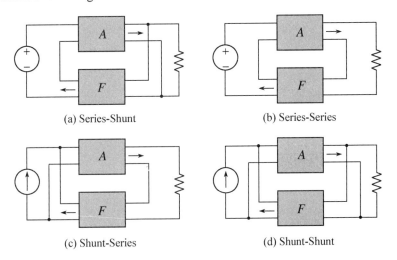

(a) Series-Shunt   (b) Series-Series

(c) Shunt-Series   (d) Shunt-Shunt

Figure 4.11　Negative Feedback Topologies

图 4.11　负反馈拓扑结构

1. Series-Shunt Configuration [Figure 4.11(a)]

Voltage is sampled at the output for feedback, and voltage is being subtracted at the input. In this configuration, feedback increases input resistance and decreases output resistance.

2. Series-Series Configuration [Figure 4.11(b)]

Current is sampled at the output, and voltage is being subtracted at the input. In this configuration, feedback increases both input and output resistance.

3. Shunt-Series Configuration [Figure 4.11(c)]

Current is sampled at the output, and current is being subtracted at the input. In this configuration, feedback decreases input resistance and increases output resistance.

4. Shunt-Shunt Configuration [Figure 4.11(d)]

Voltage is sampled at the output, and current is being subtracted at the input. In this configuration, feedback decreases both input and output resistance.

4.4.3 AC Analysis in Feedback (反馈交流分析)

Figure 4.12 shows a cascaded two-stage amplifier circuit with negative feedback.

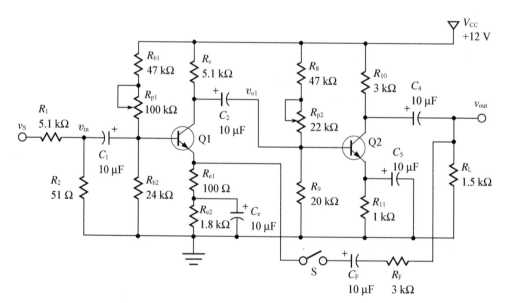

Figure 4.12 Transistor Amplifier with Feedback Circuit
图 4.12 带反馈的晶体管放大电路

Focus on the first-stage amplifier, you will find it almost the same as shown in Figure 4.2 but with a slight difference: R_{e2} is bypassed by C_e, but R_{e1} is not. This will cause the AC equivalent circuit to be different. Figure 4.13 shows the AC small-signal equivalent circuit of this part.

To calculate the voltage gain A_{v1}, notice that the current flowing through R_{e1} is $i_b + i_c$. Applying KVL to B–E–Ground loop gives

$$v_{in} = v_{BE} + v_E = r_{be} \cdot i_b + R_{e1} \cdot (i_b + i_c) \qquad (4.13)$$

When load is open $(R \rightarrow \infty)$, the voltage gain is

$$A_{v1} = \frac{v_{ol}}{v_{in}} = -\beta \frac{R_c}{r_{be} + (1 + \beta) R_{e1}} \qquad (4.14)$$

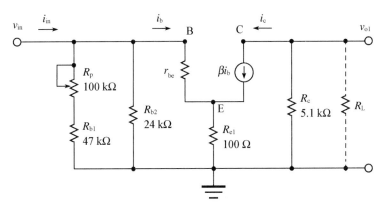

Figure 4.13　Equivalent AC Circuit Around Q1
图 4.13　Q1 附近的交流等效电路

Since β of a transistor is commonly much greater than 1, A_{v1} is approximately $-R_c / R_{e1}$. This means that the voltage gain of this negative feedback amplifier is almost not affected by the parameters of the transistor itself.

Formula (4.13) also gives input resistance as

$$r_i' = \frac{v_{in}}{i_b} = r_{be} + (1 + \beta) R_{e1} \qquad (4.15)$$

Using the method described above, you may analyze the second-stage amplifier in open loop system (R_F and C_F disconnected).

In AC closed-loop system, $R_F - R_{e1}$ loop forms series-shunt feedback circuit, which gives the feedback factor

$$F = \frac{R_{e1}}{R_F + R_{e1}}$$

hence the total gain of this amplifier $A_v \approx 1/F$.

4.5　Objectives (实验目的)

The objectives for this lab are:

1. to recognize 3 basic configurations of a transistor amplifier:

- common emitter
- common collector
- common base

2. to investigate quiescent point of an amplifier：
   - why is quiescent point important
   - how to setup the quiescent point

3. to investigate voltage gain，input and output resistance：
   - how to perform theoretical calculation
   - how to measure input and output resistance

4. to learn the frequency characteristics of an amplifier.

5. to investigate the negative feedback.

本实验研究三极管放大器的静态工作点、小信号交流电压放大倍数、输入阻抗、输出阻抗、频响特性，以及负反馈电路对这些参数的影响.

4.6　Experimental Procedure（实验步骤）

You will learn much more by actually building and analyzing real circuits. For successful circuit-building exercises，follow these steps：

1. Draw the schematic diagram for the circuit to be analyzed.

2. Mathematically analyze the circuit，solving for all voltage and current values.

3. Carefully measure and record all component values prior to circuit construction.

4. Carefully build this circuit on a breadboard or other medium. Check the accuracy of the circuit's construction，following each wire to each connection point，and verifying these elements one-by-one on the diagram.

5. Carefully measure all voltages and currents，to verify the accuracy of your analysis.

If there are any errors larger than expected，check your circuit construction against the diagram carefully，then measure and calculate the values again. To save time and to reduce the possibility of error，you should begin with a very simple circuit and incrementally add components to increase its complexity after each analysis，rather than building a whole new circuit for each practice problem.

4.6.1　Single-Stage Transistor Amplifier(单级晶体管放大器)

Quiescent Point

1. Wire the circuit as shown in Figure 4.2.

2. Turn off the signal source v_s, Connect v_{in} to ground, set collector voltage V_C to 3 V by adjusting R_{p1}. Note down the resistance value of potentiometer R_{p1} and base voltage V_B.

You must first disconnect all voltage sources from the circuit whose resistance you want to measure. That's because the multimeter will inject a known voltage into the circuit so that it can measure the current and then calculate the resistance. If there are any outside voltage sources in the circuit, the calculated resistance will be wrong. Secondly, it may be necessary to disconnect the resistor from the circuit to measure the correct resistance of the resistor. The resistance measured by a digital multimeter is the total resistance through all possible paths between the test lead probes. The resistance of all components connected in parallel with a component being tested affects the resistance reading, usually lowering it.

Input a sine wave of 10 kHz to the amplifier, note down the output voltage of maximum undistorted signal and its corresponding input magnitude.

3. Repeat last step, set V_C to 6 V and 9 V. Fill your recorded data in Table 4.2.

Table 4.2   Investigation of Quiescent Points
表 4.2   静态工作点研究

$V_C(V)$	$V_B(V)$	$R_{p1}(k\Omega)$	$I_B(\mu A)$	$I_C(mA)$	β	$v_S(V)$	$v_o(V)$	A_v
3								
6								
9								

Take the average β as a basic parameter of the transistor, calculate the voltage gain at different Q-point. You may add an extra column to Table 4.2 to list the theoretical A_v as a reference for your experimental results.

AC Performance

1. Set V_C to 6 V, plot the DC load line and mark the quiescent point on the graph.
2. Apply 10 kHz sine wave to the input, measure the voltage gain at different load, fill your recorded data in Table 4.3.
3. Based on β, which is calculated by $\beta = I_{CQ}/I_{BQ}$, calculate the voltage gain A_v of your amplifier [formula (4.8)] at different load. Compare your measurement and calculation in Table 4.3.

Table 4.3　Voltage Gain at Different Load ($f = 10$ kHz)

表 4.3　不同负载下的电压放大倍数($f = 10$ kHz)

$R_L(\text{k}\Omega)$	$v_S(\text{v})$	$v_{out}(\text{V})$	A_v (experimental)	A_v (theoretical)
2.2				
5.1				
(open)				

4. Open the resistor R_2, take R_1 as test resistor, input a sine wave of 10 kHz from function generator into the amplifier. Measure the voltage v_S and v_{in}, calculate the input resistance of the amplifier according to equation (4.9).
   $R_1 = 5.1 \text{ k}\Omega$, $v_S = \underline{\qquad}$, $v_{in} = \underline{\qquad}$, $r_i = \underline{\qquad}$.
   Watch the output signal while measuring. Since we want to measure the linear parameters of the amplifier, the output voltage of the signal source v_S should not be too large to avoid output clipping. You may add a load to decrease the output.

5. Keep the input signal unchanged while measure the output voltage with or without load. Calculate the output resistance of the amplifier according to equation (4.11).
   $v_{out} = \underline{\qquad} (R_L = \infty)$,
   $v_o = \underline{\qquad} (R_L = 5.1 \text{ k}\Omega)$,
   $r_o = \underline{\qquad}$.

Frequency Response

In this procedure, you will be applying a bunch of sinusoids at different frequencies from 50 Hz to as high as 5 MHz to measure the ratio of undistorted output amplitude to input amplitude. Fill your recorded data in Table 4.4, plot the frequency response curves in a graph. You need to use logarithmic scale for both X and Y axes in frequency response graph, and the Y axis should be scaled in dB.

Find the -3 dB points f_L (low cutoff frequency) and f_H (high cutoff frequency).

Calculate the bandwidths of the amplifier:
$R_L = 5.1 \text{ k}\Omega$, $f_L = \underline{\qquad}$ Hz, $f_H = \underline{\qquad}$ Hz, Bandwidth $= \underline{\qquad}$ Hz.
$R_L = \infty$, $f_L = \underline{\qquad}$ Hz, $f_H = \underline{\qquad}$ Hz, Bandwidth $= \underline{\qquad}$ Hz.

Table 4.4　Frequency Response of the Amplifier (Gain = 20 lg $|v_o/v_i|$)

表 4.4　放大器频率响应(Gain = 20 lg $|v_o/v_i|$)

f (Hz)	$R_L = 5.1$ kΩ			$R_L = \infty$		
	v_i(mV)	v_o(V)	Gain(dB)	v_i(mV)	v_o(V)	Gain(dB)
50						
100						
200						
500						
1 k						
2 k						
5 k						
⋮						

Harmonic Distortion

1. Apply a single frequency sinusoidal signal (10 kHz, 3 mV$_{PP}$) to the amplifier. In harmonic analysis, this frequency is called *fundamental frequency f_0*. Use the FFT function of the oscilloscope to measure the output amplitude of each harmonic frequency. Let L_1, L_2, L_3 be amplitude of f_0, $2 \times f_0$ and $3 \times f_0$ respectively.

2. Repeat last measurement by using different input amplitude. Fill your recorded data in Table 4.5. You may need formula (1.4) to calculate the n-th **harmonic distortion**.

Table 4.5　Harmonic Distortion (f_0 = 10 kHz)

表 4.5　谐波失真 (f_0 = 10 kHz)

v_i(mV)	L_1(dB)	L_2(dB)	L_3(dB)	d_2(%)	d_3(%)
3					
6					
9					
12					

4.6.2　Negative Feedback Amplifier(负反馈放大器)

Quiescent Points

1. Refer to Figure 4.12, construct a two-stage amplifier.
2. Disconnect the feedback loop R_F, C_F (open-loop). Set collector voltage of both transistors to 6 V by adjusting R_{p1} and R_{p2} separately.

Table 4.6   Quiescent Points in Feedback Circuit

表 4.6   反馈电路的静态工作点

	First-Stage				Second-Stage			
	V_{C1} (V)	V_{B1} (V)	R_{p1} (kΩ)	β_1	V_{C2} (V)	V_{B2} (V)	R_{p2} (kΩ)	β_2
open-loop	6				6			
closed-loop								

3. Connect feedback loop (closed-loop) while keeping R_{p1} and R_{p2} unchanged, check if the quiescent points are changed. Fill your recorded data in Table 4.6. Calculate the β of the two transistors.

Since the voltage gain will be very large. You need to ground input to avoid interference to the quiescent point during measurement.

AC Performance

1. Disconnect feedback loop and coupling capacitor C_2, measure the voltage gain of the first-stage amplifier. Compare your result with Table 4.3. Why are they so different? Please show your theoretical analysis.

2. Connect the capacitor C_2, keep the feedback loop open. Apply 10 kHz sine wave to the input, measure the output signal of each stage. Let v_{o1} be the 1st stage's output voltage, v_{out} be the 2nd stage's output voltage. Calculate the voltage gain of each stage and total gain of the amplifier.

Table 4.7   AC Voltage Gain of the Amplifier ($f = 10$ kHz)

表 4.7   放大器交流电压增益($f = 10$ kHz)

	v_i (mV)	v_{o1} (mV)	v_{out} (V)	A_{v1}	A_{v2}	A_v
open-loop	1					
	2					
closed-loop	2					
	10					
	20					

3. Connect feedback loop (closed-loop), repeat last step. Fill your recorded data in Table 4.7.

4. Measure the input and output resistance for both open-loop and close-loop circuit. Note down your measurement and calculation in Table 4.8.

Table 4.8 Input and Output Resistance in Feedback Amplifiers

表 4.8 反馈放大器的输入和输出电阻

$f = 10$ kHz	Input Resistance ($R_S = 5.1$ kΩ)			Output Resistance ($R_L = 1.5$ kΩ)		
	v_S(mV)	v_i(mV)	r_i(kΩ)	v_{out}(V)	v_o(V)	r_o(kΩ)
open-loop						
closed-loop						

Frequency Response of a Negative Feedback Amplifier

Apply sinusoids at different frequencies to the amplifier, measure the ratio of **undistorted output** amplitude to input amplitude. Fill your recorded data in Table 4.9, plot the frequency response curves in a graph.

Table 4.9 Frequency Response of the Amplifier

表 4.9 放大器频率响应

f (Hz)	Open-Loop			Closed-Loop		
	v_i(mV)	v_o(V)	A_v(dB)	v_i(mV)	v_o(V)	A_v(dB)
50						
100						
200						
500						
1 k						
2 k						
5 k						
⋮						

Harmonic Distortion

Apply sinusoidal signal of 10 kHz to the amplifier. Use the FFT function of the oscilloscope to measure the output amplitude of each harmonic frequency. Let L_1, L_2, L_3 be amplitude of fundamental frequency f_0, second harmonic $2 \times f_0$ and third harmonic $3 \times f_0$ respectively. For both open-loop and closed-loop condition, write the amplitude of harmonic frequencies in Table 4.10.

Table 4.10   Harmonic Distortion ($f_0 = 10$ kHz, $v_{out} = 1$ V$_{PP}$, $R_L = 1.5$ kΩ)

表 4.10   谐波失真($f_0 = 10$ kHz, $v_{out} = 1$ V$_{PP}$, $R_L = 1.5$ kΩ)

	v_i(mV)	L_1(dB)	L_2(dB)	L_3(dB)	d_2(%)	d_3(%)
open-loop						
closed-loop						

In order to make a reasonable comparison, the measurement should be taken under the same output voltage level for both open-loop and closed-loop.

4.7 Questions(问题)

1. Fundamentally, an amplifier is a device that takes in a low-power signal and gives a magnified output signal. Explain how it is possible for such a device to exist. Doesn't the Law of Energy Conservation preclude the existence of a power-boosting device?

2. Investigate the circuit in Figure 4.2, which element is the **predominant** factor in determining the low-frequency response for the amplifier. How will you lower the low cutoff frequency?

【**Vocabulary**】

emitter：发射极

collector：集电极

base：基极

common-emitter configuration：共射极配置(电路)

schematic：电路原理图

voltage divider：分压器

biasing：偏置

bypass capacitor：旁路电容

distortion：失真

quiescent point：静态工作点 (常简写为 Q-point)

small signal model：小信号模型

input impedance：输入阻抗

output impedance：输出阻抗

frequency response：频率响应

amplitude response：幅度响应

phase response：相位响应

steady-state sinusoidal：稳态正弦

cutoff frequency：截止频率

open-loop gain：开环增益

closed-loop gain：闭环增益

series-shunt：电压-串联反馈结构

series-series：电流-串联反馈结构

shunt-series：电流-并联反馈结构

shunt-shunt：电压-并联反馈结构

[Translations]

1. Depend on which terminal is "common", three basic configurations are available：common-emitter, common-collector and common-base. The word "commons" here means "AC-grounded".

   根据公共端的选取,(三极管放大器)有三种基本配置形式:共射极、共集电极和共基极。此处"共"的意思是"交流接地".(电子线路中,地线常被作为系统内、系统与系统之间的公共端)

2. Recall that a capacitor acts like a short circuit for AC signals, but like an open circuit for DC, this allows the input AC signal to pass through the capacitor while stopping DC voltage from the power supply from leaking out. This is called AC coupling the input.

   别忘了,电容对交流如同短路,而对直流如同断路,这个特性允许输入交流信号通过,同时阻止来自电源的直流电压漏过(而影响输入信号源).此谓输入交流耦合.

3. Knowing that the base-emitter junction of NPN transistors needs to be forward biased for the transistor to conduct, if we want our amplifier to amplify signals all the time, the BE junction needs to be forward biased all the time. This means that we must have $V_{BE} > 0.7\,\text{V}$ (for silicon transistor) all the time.

   我们知道,NPN型三极管的基极和发射极需要保持正向偏置,以使其导通.如果我们要让放大器始终都能放大信号,BE结就要始终保持正向偏置.这意味着我们必须让 V_{BE} 始终大于 0.7 V(对于硅管来说).

4. However, if we add some DC offset to the base voltage, such that even with the swing of our input voltage $V_{BE} > 0.7\,\text{V}$ all the time, then we can guarantee that our transistor will be always on. This is called biasing the transistor.

   然而,如果我们给基极电压加一点直流偏置,使得即使有(输入)摆动,输入电压 V_{BE} 也始终大于 0.7 V,我们就可以确保晶体管总是导通的.这就是给晶体管设置偏置(电路).

5. The quiescent point of this amplifier is somewhere along this slope. We now need to find out which curve we need to use. To select a suitable curve, we need to calculate the base current at the quiescent point I_{BQ}.

这个放大器的静态工作点位于这条斜线上某个位置. 我们现在需要找出使用哪条曲线(和这条直线相交的交点). 要选择合适的曲线,我们需要计算静态工作点的基极电流 I_{BQ}.

6. Where thermal voltage V_T is a temperature-related variable. At approximately room temperature (22 ℃ or 295 K), V_T is about 25 mV.

这里的热电压 V_T 是一个与温度相关的变量. 在室温下(22 ℃,295 K),V_T 大约是 25 mV.

7. Input impedance of a circuit is the resistance seen by the input signal when going through the circuit. Output impedance is the impedance seen when looking into the output port of the circuit.

电路的输入阻抗是当输入信号进入这个电路时,从输入信号的角度看到的阻抗. 输出阻抗是从输出端看到的这个电路的阻抗.

8. For frequencies in the low to mid-range, the input impedance of a BJT transistor amplifier is purely resistive.

对于低频到中频范围,BJT 三极管放大器的输入阻抗是纯阻性的.

9. A high input resistance means the circuit will draw less current from the previous stage. The higher the input resistance of a circuit is, the less impact it has on the preceding circuit. The reverse is also true.

高输入电阻意味着这个电路从前级电路吸纳较少的电流. 一个电路的输入电阻越高,它对前级电路的影响越小. 反之亦然.

10. The steady-state sinusoidal frequency response of a circuit is described by the phasor transfer function. A bode plot is a graph of the magnitude (in dB) or phase of the transfer function with respect to frequency as shown in Figure 4.9.

电路对于稳态正弦信号的频率响应通过相量传递函数表述. 波特图是传递函数的幅度(以 dB 为单位)或相位关于频率的曲线图,如图 4.9 所示.

11. In most electronic circuits, the word "bandwidth" refers to the range of frequencies over which the system produces a specified level of performance. In the case of frequency response, bandwidth in Hz refers to the range of frequencies more than 3 dB below the maximum amplitude gain, or 3 dB bandwidth.

在大多数电子线路中,"带宽"一词指的是系统产生给定电平性能的频率范围. 对于频率响应来说,以 Hz 为单位的带宽指(达到)最大幅度增益以下 3 dB 的频率范围,或称"3 dB 带宽".

12. Feedback is the process by which a fraction of the output signal，either a voltage or a current，is used as an input. If the feedback fraction is opposite in value or phase（"anti-phase"）to the input signal，then the feedback is said to be negative feedback.

反馈是这样的过程：它将输出信号的一部分，无论是电压还是电流，用作输入. 如果反馈的部分在数值或相位与输入信号相反，则该反馈被称为负反馈.

13. Suppose the feedback loop applies a fraction F of the output v_o to one of the subtractor inputs so that it subtracts from the circuit input v_i applied to the other subtractor input. The result of subtraction applied to the amplifier input is $v_i' = v_i - F \cdot v_o$.

假设反馈回路把输出的一部分 Fv_o 加载到减法器的一个输入端，让它从减法器另一端输入的电路输入 v_i 中减去. 加在放大器输入端的相减结果是 $v_i' = v_i - F \cdot v_o$.

14. The feedback topology depends on whether we amplify voltage or current，and on whether we are sampling voltage or current at the output for feedback use. There are four different configurations.

反馈拓扑结构取决于我们是放大电压还是电流，以及在输出端采电压还是采电流用于反馈，有四种不同的（反馈结构）配置.

15. Secondly，it may be necessary to disconnect the resistor from the circuit to measure the correct resistance of the resistor. The resistance measured by a digital multimeter is the total resistance through all possible paths between the test lead probes. The resistance of all components connected in parallel with a component being tested affects the resistance reading，usually lowering it.

第二，有必要将电阻从电路中断开，以便测量正确的阻值. 由数字万用表测得的电阻是表笔之间所有可能的通路上的电阻的总值. 与被测元件并联的所有元件会影响电阻的（正确）读数，通常会使这个读数变小.

16. Focus on the first-stage amplifier，you will find it almost the same as shown in Figure 4.2 but with a slight difference：R_{e2} is bypassed by C_e，but R_{e1} is not. This will cause the AC equivalent circuit to be different.

关注第一级放大器，你会发现它与图 4.2 几乎一样，只有很小的差别：R_{e2} 被 C_e 旁路，而 R_{e1} 没有. 这将导致交流等效电路的不同.

17. Since β of a transistor is commonly much greater than 1，A_{v1} is approximately $-R_c/R_{e1}$. This means that the voltage gain of this negative feedback amplifier is almost not affected by the parameters of the transistor itself.

由于晶体管的 β 值通常远大于 1，A_{v1} 近似等于 $-R_c/R_{e1}$. 这意味着这个负反馈放大器的电压放大倍数几乎不受三极管自身参数的影响.

5
Differential Amplifier
差动放大器

A **differential amplifier** is a type of electronic amplifier that amplifies the difference between two input voltages but suppresses any voltage common to the two inputs. Differential amplifiers are used to amplify balanced differential signals，which are commonly used to communicate small signals in electrically noisy environments.

5.1 Introduction（简介）

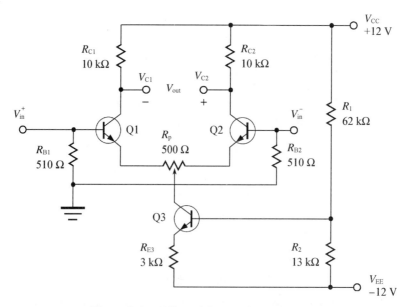

Figure 5.1 Differential Amplifier Using BJT

图 5.1 结型晶体管构成的差动放大器

The differential amplifier using two transistors Q1 and Q2 is designed as shown in Figure 5.1. The third transistor Q3 acts as a constant current source. The potentiometer R_p is used to adjust quiescent of the transistors to make $V_{out} = 0$ when $V_{in}^+ = V_{in}^- = 0$.

This circuit has a unique topology: two inputs and two outputs, although you can tap the signal from one output only. The input signals, V_{in}^+ and V_{in}^-, are applied to the base of the transistors while the output is collected across their collector terminals V_{C1} and V_{C2}. The circuit can amplify the difference between two input voltages but suppress any voltage common to the two inputs.

5.1.1 Gain and Rejection (增益与抑制)

How does this amplifier amplify differential signals and reject common ones? The bias condition assumes equal voltages at V_{B1} and V_{B2}, forcing the bias current I_E to split equally between the transistors resulting in $I_{C1} = I_{C2}$. With $R_{C1} = R_{C2}$, equal voltages develop at V_{C1} and V_{C2}.

Suppose a differential signal is applied to the inputs. This will change the base voltages to $V_{B1} + \Delta V$ and $V_{B2} - \Delta V$. As Q1 conducts a little more and Q2 a little less, I_E now splits unevenly creating $I_{C1} > I_{C2}$. This decreases V_{C1} and increases V_{C2}. Thus the voltage changes at each output due to a differential input.

Now suppose a common-mode input signal is applied, increase both inputs to $V_{B1} + \Delta V$ and $V_{B2} + \Delta V$, the conduction level of neither transistors changes (both bases and emitters move towards the same direction), the collector currents do not change. $I_{C1} = I_{C2} \approx I_E/2$. Subsequently, the voltages at V_{C1} and V_{C2} remain the same.

Actually, asymmetrical properties of the circuit may impact $I_{C1} = I_{C2}$, slightly shifting the levels at V_{C1} and V_{C2}. The rejection is not perfect. However, it can still be effective at removing a large part of common-mode noise or a DC bias common to both inputs.

5.1.2 Common-Mode Rejection (共模抑制)

The output of the amplifier is ideally proportional to the difference between the two input voltages

$$V_{out} = A_d (V_{in}^+ - V_{in}^+)$$

where A_d is the differential gain of the amplifier.

In fact, however, the gain is not quite equal for the two inputs. This means, for instance, that even if V_{in}^+ and V_{in}^- are equal, the output will not be zero, as it would be in the ideal case. Thus, a more realistic expression for the output of a differential amplifier includes a second term

$$V_{out} = A_d(V_{in}^+ - V_{in}^-) + A_c \frac{V_{in}^+ + V_{in}^-}{2} \tag{5.1}$$

where A_c is the common-mode gain of the amplifier.

As differential amplifiers are often used to suppress noise or bias voltages at both inputs, a lower common-mode gain is usually desired. The **common-mode rejection ratio** (CMRR), usually defined as the ratio between **differential-mode gain** and **common-mode gain**, indicates the ability of the amplifier to eliminate signals that are common to both inputs. The CMRR is often expressed in dB and is defined as

$$CMRR = 20 \lg \left| \frac{A_d}{A_c} \right| \tag{5.2}$$

In a perfectly symmetric differential amplifier, A_c is zero, and the CMRR is infinite.

5.1.3 Differential Gain (差模增益)

Take Q1 and Q2 as current sources controlled by their base voltages. The small signal collector current can be expressed as

$$i_C = g_m \cdot v_{BE} \tag{5.3}$$

where the **transconductance** g_m is set by the DC collector current. $g_m \approx I_C/V_T$, where the thermal voltage V_T is about 25 mV at room temperature.

Then, the collector resistor R_C transforms small signal current i_C back to voltage

$$v_C = -R_C \cdot g_m \cdot v_{BE}$$

Taking the input v_S into account, notice that it is divided equally across each base-emitter junction, but with opposite polarities. When the emitter coupling resistor R_p is not present, we get a single-ended output voltage for each transistor,

$$v_{C1} = -v_S/2 \cdot R_{C1} \cdot g_m$$
$$v_{C2} = +v_S/2 \cdot R_{C2} \cdot g_m \tag{5.4}$$

Subtracting the two outputs gives the differential output,

$$v_{out} = v_{C1} - v_{C2} = -v_S \cdot R_C \cdot g_m \tag{5.5}$$

Q3 sets the bias current at $I_{C3} \approx I_{E3} \approx (V_{B3} - 0.7\text{ V} - V_{EE})/R_{E3} = 1.15$ mA, which divides equally between Q1 and Q2 giving $I_{C1} = I_{C2} \approx I_{C3}/2 = 0.575$ mA. Therefore

◀◀◀

$$g_m = I_C / V_T = 0.023 \text{ A/V}$$

Finally, The single-ended gain is calculated as

$$v_{C1} / v_S = R_{C1} \cdot g_m/2 = 115$$

Take the coupling resistor R_p into consideration. Assume the two base voltages are $V_{B1} + \Delta V_B$ and $V_{B2} - \Delta V_B$, where $\Delta V_B = v_S/2$. These change emitter voltages to $V_{E1} + \Delta V_E$ and $V_{E2} - \Delta V_E$, and eventually change the collect current i_C flowing through R_C:

$$i_C = \frac{\beta}{\beta+1} i_E = \frac{\beta}{\beta+1} \frac{\Delta V_E}{R_p/2} \tag{5.6}$$

hence

$$\begin{aligned} v_C &= i_C \cdot R_C \\ &= \frac{\beta}{\beta+1} \cdot \frac{\Delta V_E}{R_p/2} \cdot R_C \end{aligned} \tag{5.7}$$

where β is the base-collector current gain of the transistor.

Recall $\Delta V_B - \Delta V_E = v_{BE} = i_B r_{be}$. Putting it all together, we get a single-ended voltage gain for transistor Q1:

$$\begin{aligned} A_{d1} &= \frac{v_{C1}}{\Delta V_B} \\ &= \frac{\beta R_{C1}}{r_{be} + (\beta+1) R_p/2} \end{aligned} \tag{5.8}$$

Since $\beta \gg 1$, the differential-mode gain A_d is approximately $\dfrac{2R_C}{R_p}$.

Take r_{be} into account,

$$r_{be} = r_{bb'} + \beta \frac{V_T}{I_C} \tag{5.9}$$

then,

$$A_d \approx \frac{R_C}{1/g_m + R_p/2} \tag{5.10}$$

5.2　Configurations（配置形式）

Figure 5.1 shows the typical configuration that differential amplifier can be used. Besides, differential amplifier has other operation modes, each with its individual characteristics.

5.2.1 Single-Ended Input Mode (单端输入方式)

The differential pair can be used as an amplifier with a single-ended input and dual-ended output if one of the inputs is grounded or fixed to a reference voltage. We can construct it by connecting one of two input pins with input signal V_{in} and the other one connect to ground as shown in Figure 5.2. There are two cases for output signal according to its polarity with the input signal, inverting output and non-inverting output.

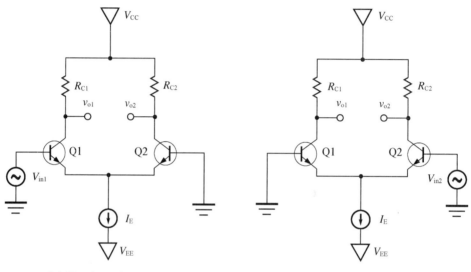

(a) Non-inverting Output   (b) Inverting Output

Figure 5.2   Single-Ended Input Mode

图 5.2   单端输入模式

5.2.2 Single-Ended Output Mode (单端输出方式)

If the differential output is not desired, then only one output can be used, disregarding the other output. This configuration is referred to as single-ended output. The gain is half that of the stage with differential output.

5.2.3 Common Input Mode (共模输入方式)

The differential amplifier has the same input signal ($V_{in1} = V_{in2}$) as shown in Figure 5.3. In ideal case, the output voltage will be zero, but actually there will always be very low voltage measured due to noise and asymmetrical properties.

5.3   Objective (实验目的)

In this lab, you will be investigating differential amplifiers:

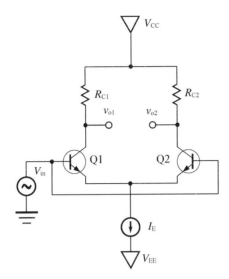

Figure 5.3   Common Input Mode Differential Amplifier
图 5.3   差动放大器共模输入

- to understand the DC and AC operation of a differential amplifier;
- to measure DC voltages and currents in differential amplifier;
- to obtain measured values of differential-mode gain A_d and common-mode gain V_c, to investigate common-mode rejection ratio CMRR.

5.4   Experimental Procedure (实验步骤)

1. Quiescent setting.

Construct the circuit shown in Figure 5.1, then make the circuit quiescent (no signal applied) by connecting both bases to ground, i.e. $V_{in}^+ = V_{in}^- = 0$.

Adjust R_p so that $V_{C1} = V_{C2}$, measure the DC voltages and fill the recorded data in Table 5.1. The closer V_{C1} and V_{C2} are, the higher the CMRR is.

Table 5.1   Setup Quiescent ($V_{B1} = V_{B2} = 0$ V)
表 5.1   设置静态工作点 ($V_{B1} = V_{B2} = 0$ V)

DC values	V_{C1}	V_{C2}	V_{E1}	V_{E2}	V_{B3}	V_{C3}	V_{E3}
Measured (V)							
Calculated (V)							

2. Measure differential-mode gain.

Set differential amplifier to dual-ended input mode. Apply 0.1 V DC to V_{in}^+, -0.1 V DC to V_{in}^-. Because the input resistance of the amplifier is less than 510 Ω (R_{B1} or R_{B2}), not large enough relative to the output resistance of DC source, you need to adjust the source carefully to ensure ± 0.1 V at the input terminals.

Write your measured data in Table 5.2. Calculate the differential-mode gain of the amplifier.

Table 5.2   Differential-Mode Gain
表 5.2   差模增益

V_{in}^+ (V)	V_{in}^- (V)	Measured		Calculated		
		v_{C1} (V)	v_{C2} (V)	A_{d1}	A_{d2}	A_d
0.1	-0.1					

The differential-mode gain is calculated as:

$$A_{d1} = \frac{v_{C1} - V_{C1}}{V_{in}^+ - V_{in}^-} \tag{5.11}$$

where V_{C1} is the quiescent voltage measured in Table 5.1.

3. Measure common-mode gain and CMRR.

Set differential amplifier to common input mode. Apply $+0.1$ V and -0.1 V to the input. Write the measured data in Table 5.3.

Table 5.3   Common-Mode Gain
表 5.3   共模增益

V_{in} (V)	Measured		Calculated		
	v_{C1} (V)	v_{C2} (V)	A_{c1}	A_{c2}	CMRR
$+0.1$					
-0.1					

The common-mode gain is calculated as

$$A_{c1} = \frac{v_{C1} - V_{C1}}{V_{in}} \tag{5.12}$$

Based on A_d and A_c you have measured, the common-mode rejection ratio (CMRR) can be calculated by using equation (5.2).

4. Measure differential-mode gain in single-ended input mode.

Ground V_{in}^-, apply ± 0.1 V DC and 50 mV$_{PP}$/1 kHz sine wave to V_{in}^+ respectively.

Measure the outputs of the amplifier and write them in the Table 5.4. calculate the differential-mode gain A_d.

Table 5.4 Single-ended Input Mode

表 5.4　单端输入方式

V_{in}(V)	v_{C1}(V)	v_{C2}(V)	v_o	A_d
+0.1 V DC				
−0.1 V DC				
50 mV$_{PP}$/1 kHz				

Answer following questions based on your experiment results：

1. How to improve common-mode rejection，i. e. to increase CMRR？

2. How do the resistors R_{B1} and R_{B2} affect the circuit? Will the CMRR be greatly changed if remove them?

［实验简介］

　　差动放大器又称差分放大器,是一种将两个输入端电压的差以一固定增益放大的电子电路.它对差模信号可以起到有效的放大作用,同时由于电路的对称性,能够有效地减小共模干扰,以及抑制由于电源波动或温度变化而引起的零点漂移,因而获得广泛的应用.集成运算放大器的输入级大都为差动放大器.它也常被用作多级放大器的前置级,因为前置放大器的信号源往往是高内阻电压源,它要求具有足够高的输入电阻,这样才能有效地接收到信号.

　　典型的晶体管差动放大电路由两个发射极耦合的对称三极管构成.本实验研究这种电路的特性,它的差模放大倍数、共模放大倍数,以及共模抑制比.

［Vocabulary］

　　differential amplifier：差分放大器,差动放大器

　　suppress：抑制

　　common-mode input：共模输入

　　differential-mode gain：差模增益

　　common-mode rejection ratio：共模抑制比（CMRR）

　　polarity：极性

　　transconductance：跨导

　　single-ended：单端的

　　dual-ended：双端的

［Translations］

1. The bias condition assumes equal voltages at V_{B1} and V_{B2}，forcing the bias current I_E to split equally between the transistors resulting in $I_{C1} = I_{C2}$. With

$R_{C1} = R_{C2}$, equal voltages develop at V_{C1} and V_{C2}.

偏置条件假设满足在 V_{B1} 和 V_{B2} 两处电压相等,由此迫使偏置电流 I_E 在两个晶体管之间均分,从而 $I_{C1} = I_{C2}$. 因 $R_{C1} = R_{C2}$, V_{C1} 和 V_{C2} 两处电压等量变化.

2. Suppose a differential signal is applied to the inputs. This will change the base voltages to $V_{B1} + \Delta V$ and $V_{B2} - \Delta V$. As Q1 conducts a little more and Q2 a little less, I_E now splits unevenly creating $I_{C1} > I_{C2}$.

假设差分信号加在(两个)输入端.这将使基极电压变成 $V_{B1} + \Delta V$ 和 $V_{B2} - \Delta V$. 由于 Q1 导通多一点,Q2 导通少一点,此时 I_E 不均分,导致 $I_{C1} > I_{C2}$.

3. In fact, however, the gain is not quite equal for the two inputs. This means, for instance, that even if V_{in}^+ and V_{in}^- are equal, the output will not be zero, as it would be in the ideal case. Thus, a more realistic expression for the output of a differential amplifier includes a second term.

然而事实上,对于两个输入端的增益并不完全相等.这意味着,例如,即使 V_{in}^+ 和 V_{in}^- 相等,输出也不是如理想情况那样等于零.因此,差动放大器输出的一个更为实际的表达式包含第二项.

4. As differential amplifiers are often used to suppress noise or bias voltages at both inputs, a lower common-mode gain is usually desired. The common-mode rejection ratio (CMRR), usually defined as the ratio between differential-mode gain and common-mode gain, indicates the ability of the amplifier to eliminate signals that are common to both inputs.

由于差动放大器常被用于抑制(同时加在)两个输入端的噪声或偏置电压,通常希望有较低的共模增益.共模抑制比(CMRR),通常被定义为差模增益与共模增益之比,表征了放大器对于两个输入端的相同信号的抵消能力.

5. Taking the input v_S into account, notice that it divides equally across each base-emitter junction, but with opposite polarities. Putting it all together we get a single-ended output voltage for each transistor ...

考虑 v_S,注意到它在每个发射结等分,但极性相反.综上,我们得到每个晶体管的单端输出电压为……

6. The differential pair can be used as an amplifier with a single-ended input and dual-ended output if one of the inputs is grounded or fixed to a reference voltage.

如果其中一个输入端接地,或者固定在一个参考电压上,差分对可用作单端输入、双端输出的放大器.

6 Operational Amplifier Basics
运算放大器入门

6.1 Introduction (简介)

An **operational amplifier** or op-amp is a high gain DC amplifier. It is very widely used in both analog and digital design because its operating characteristics can be significantly changed by connecting different external components. The operational portion of the name derives from its original application for performing mathematical operations in analog computers.

6.1.1 Notation and Terminology(符号和术语)

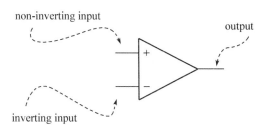

Figure 6.1　A Diagram of Op-amp
图 6.1　运算放大器的框图

The operational amplifier schematic symbol is a triangle and with at least two inputs and one output as shown in Figure 6.1. Note that the two inputs are marked and behave differently.

- The " + " input is the **non-inverting** or positive input of the op-amp. A signal applied to the non-inverting input is amplified by the op-amp.
- The " − " input is the **inverting** or negative input of the op-amp. A signal applied to the inverting input is amplified and inverted by the op-amp.

▶▶▶

The inputs are not positional. Either input (inverting or non-inverting) can be the top or bottom input.

6.1.2 Ideal Op-amps (理想运放)

An operational amplifier is a **direct current coupled** voltage amplifier. That is, it increases the input voltage that passes through it. The input resistance of an op-amp should be high whereas the output resistance should be low. An op-amp should also have very high open loop gain. An ideal op-amp is defined as a differential amplifier with infinite open loop gain, infinite input resistance and zero output resistance. It also has an infinite bandwidth of operation and zero offset voltage.

Although this kind of amplifier doesn't exist, the following two key points are still applicable in most **negative feedback** cases:

1. An op-amp has zero input current. Since the input resistance is very high, current flowing into the amplifier is negligible.
2. Non-inverting and inverting inputs of an op-amp are **virtual short-circuited**, as they have almost the same voltage. When one of the input terminal is grounded to GND, the other terminal is sometimes called **virtual ground**.

These concepts are helpful for mathematical analysis of the op-amp circuit.

6.1.3 Differential Amplifiers(差动放大器)

Operational amplifiers are **differential amplifiers**. They amplify the difference in voltage between the signals at the two inputs (inverting input and non-inverting input). Therefore, if you shorted the inverting and non-inverting inputs together to ground, you would (ideally) expect an output of zero volts. In fact, an actual op-amp will probably have a slight offset at one of the inputs resulting in a non-zero output. This **offset voltage** will tend to introduce slight errors in an op-amp circuit.

To allow for compensation of the offset voltage, some op-amps have two inputs that are called **offset null** inputs. If the circuit requires offset null inputs, the schematic will be as in Figure 6.2.

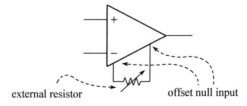

external resistor   offset null input

Figure 6.2 A Differential Amplifier with Offset Null Compensation
图 6.2 带有零漂移补偿的差动放大器

If it is necessary to compensate for the offset voltage, short the inverting and non-inverting inputs together to ground and adjust the variable resistor until the output is zero. Then reconfigure the circuit as per the schematic leaving the variable resistor in the circuit as adjusted.

Typically, we will not use the offset null inputs.

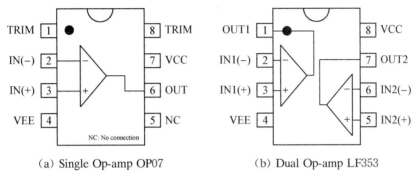

(a) Single Op-amp OP07   (b) Dual Op-amp LF353

Figure 6.3 Pin Configurations of Op-amp ICs (Top View)

图 6.3 集成运放引脚连接（顶视图）

In the following experiments, we will use OP07 or LF353 operational amplifier to build circuits. Their pin configurations are shown in Figure 6.3. Please read the data sheet, locate the inverting and non-inverting input pins and the output pin on the chip. Sketch the pin-out for the chips before construction.

6.1.4 Power Supply (电源供电)

The inverting and non-inverting input voltages (required) control the operation of the op-amp. The offset null input voltages (optional) are used to fine tune the op-amp's internal circuitry for sensitive applications.

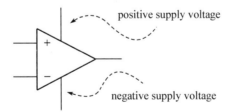

Figure 6.4 Voltage Supply for Op-amp IC

图 6.4 运算放大器供电

In addition, an op-amp IC has two inputs for power. These inputs are typically referred to as **supply voltages** or the **rails**. Typically, wiring diagrams will show the supply voltages and schematics will not show the supply voltages. **The supply voltages are always required when wiring up an op-amp even if not shown on the schematic.**

These inputs power the IC and limit the behavior of the op-amp: the inverting and non-inverting inputs should never run outside the supply voltage range, and the output voltage will never exceed the supply voltages.

It is important to note that the supply voltages are indicated as **positive** and **negative**. The positive voltage supply must be greater than the negative voltage supply. Any op-amp can be used in either single supply or dual supply configuration, but there is a difference in how each will perform.

⚠ The power supply to the operational amplifier should never be reversed in polarity. The input voltage at the positive supply pin must be greater than the input voltage at the negative supply pin. If polarity is reversed to the IC, the internal conductors can fuse and destroy the chip.

One of the most popular configurations for the supply voltages is dual supply, in which the voltages are the same value but opposite polarity. Another typical configuration is single supply. In single supply configuration, one voltage is ground and the other is a positive or negative voltage. Usually, we use positive voltage in single supply.

In this and following experiments, we use op-amp OP07. Please read the OP07's data sheet, locate the supply voltage pins on the chip. What is the polarity of the pins? How did you determine that from the data sheet (i.e. what notation did the data sheet use to indicate polarity)?

If the op-amp is configured using dual supply and the positive supply is $+12$ V, what is the voltage on the negative supply? If the op-amp is configured using single supply and the positive supply is $+12$ V, what is the voltage on the negative supply?

6.2 Objectives (实验目的)

An operational amplifier or op-amp is a high gain DC *amplifier* typically used in a closed-loop configuration. The objectives for this lab are:
1. to study op-amp notation and terminology (研究运算放大器的符号和术语)
   - op-amp terminology (运算放大器的术语)
   - schematic notation versus actual wiring practice (通过实际连线操作了解原理图的符号)
   - labeling op-amp's inputs and output (标出运算放大器的输入和输出)
2. to investigate the behavior of an op-amp in an open-loop configuration—the

comparator（研究运算放大器开环结构——比较器的工作状态）
- operation of inverting and non-inverting inputs（反相和同相输入的操作）
- impact of op-amp gain（运算放大器增益的影响）
- limitations imposed by supply voltages（受电源电压的限制）

3. to investigate the behavior of an op-amp in closed-loop configurations（研究运算放大器闭环结构）
   - inverting amplifier，non-inverting amplifier，differential amplifier（反相放大器，同相放大器，差动放大器）
   - how to adjust gain（如何调节增益）
   - how to design an amplifier with a specified voltage gain（如何设计一个给定增益的放大器）
   - to investigate the frequency characteristics of an amplifier（研究放大器的频率特性）
   - to learn to measure the input impedance and output impedance（学习测量放大器的输入阻抗和输出阻抗）

6.3　Equipments and Materials(实验器材)

- Bench supply，oscilloscope，function generator，multimeter.
- Breadboard，potentiometers，resistors，LEDs①，wires.
- For open-loop circuit，inverting and non-inverting amplifier use the OP07.

6.4　Experimental Procedure（实验步骤）

6.4.1　Voltage Comparator（电压比较器）

Operational amplifiers are differential amplifiers that amplify the difference in voltage between the two input connections，

$$V_{out} = A_v (V_+ - V_-) \tag{6.1}$$

where A_v is the voltage gain.

The comparator is in an open-loop configuration. The term **open-loop** comes from control system theory. There is no feedback in the open-loop mode of a control system. Except comparator，most of the remaining op-amp circuits you will

----

① Light emitting diode，commonly called LED，is a semiconductor element that emits light when current flows through it.

▶ ▶ ▶

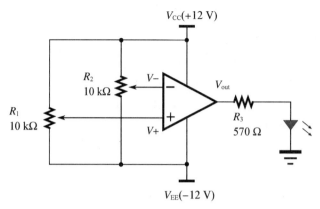

Figure 6.5　An Open-Loop Comparator

图 6.5　一个开环的比较器

be investigating are closed-loop. Operational amplifiers are usually used in closed-loop mode where there is feedback from the output back to the input.

Construct an open-loop comparator circuit shown in Figure 6.5 using OP07. Use ± 12 V for the supply voltages (note the above caution about polarity). Use two potentiometers R_1 and R_2 to set input voltage of amplifier V_+ and V_-. Monitor V_{out} with an LED. Note that LED is on only when V_{out} is greater than 0 V.

 Never apply inputs before establishing supply voltages.

1. Set V_+ to 1 V, and V_- to 2 V by adjusting R_1 and R_2. Note V_{out} and the status of LED.
2. Increase V_+ slowly to 3 V by adjusting potentiometer R_1 while watching LED. Keep R_2 unchanged (thus V_- unchanged during this procedure).
   If you notice anything happen to LED, record what happens and the value of V_+.
3. Now leave V_+ at 3 V, and increase V_- slowly to 4 V while watching LED.
   If you notice anything happen to LED, record what happens and the value of V_-.
4. Repeat the pattern of the last two steps, leapfrogging V_+ and V_- until they reach the positive supply voltage. **What pattern did you observe in LED?**

Table 6.1 Comparator Input and Output

表 6.1 比较器的输入与输出

V_+ (V)	V_- (V)	LED on/off	V_{out}
1	2		
	2		
3			
	4		
5			
...	...		
...	...		

5. Recall equation (6.1), what output would you expect if you had unity voltage gain, i.e. $A_v = 1$? Explain it.

Knowing this, what can you say about A_v for this op-amp, e.g. is it much less than 1, is it less than 1, equal to 1, greater than 1, much greater than 1? Explain your reasoning.

6.4.2 Voltage Follower (Voltage Buffer) (电压跟随器/缓冲器)

The voltage follower (or buffer) is a very simple circuit. Since $V_+ = V_-$ (virtual short) and $V_- = V_{out}$, this gives $V_{out} = V_{in}$. In other words, the voltage of output follows input (Figure 6.6).

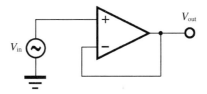

Figure 6.6 A Voltage Follower Circuit

图 6.6 电压跟随器

6.4.3 Inverting Amplifier (反相放大器)

An **inverting amplifier** circuit produces an amplified output signal that is $180°$ out of phase with the input signal. Use an OP07 to construct the closed-loop inverting amplifier circuit as shown in Figure 6.7.

1. Use ± 12 V for the supply voltages. Use a feedback resistor R_f, of 10 kΩ and an input resistor R_{in}, of 1 kΩ. When you sketch the circuit in your lab notebook, note the feedback and input resistor values.

2. Drive the amplifier with a 10 kHz sine wave with a DC offset of zero,

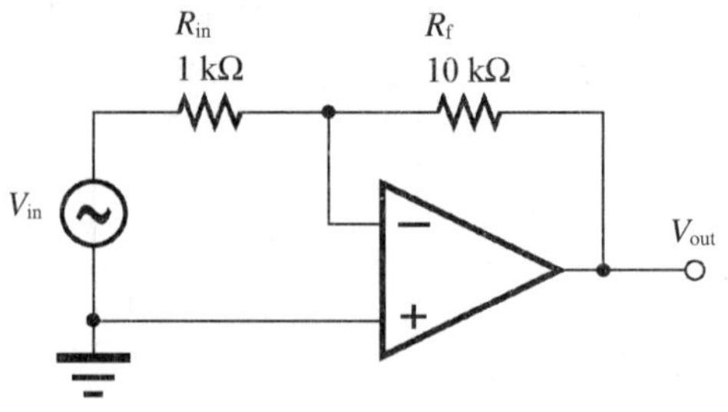

Figure 6.7　An Inverting Amplifier Circuit
图 6.7　反相放大器

starting with a small amplitude (0.1 V_{PP}). Watch both input and output signals on the oscilloscope as the amplitude increases. Sketch input and output waveform, including measurements of amplitude.

At what point does the output signal saturate (output signal will no longer look like a sine wave)? For the remainder of this exercise, use a reasonable amplitude (not too small but do not saturate the output).

A very important parameter of operational amplifier performance is **slew rate**. Slew rate is defined as the maximum rate of output voltage change over time ($\frac{dv}{dt}$) that an op-amp can output and is given units of V/μs. Slew rate is measured by applying a large signal step, such as 1 V, to the input of the op-amp, and measuring the rate of change from 10% to 90% of the output signal's amplitude. Amplifiers with higher slew rate tend to have higher bandwidth.

Slew rate is a different specification than small signal bandwidth. Giving a large magnitude of sine wave input, the output will look more like a triangle wave than a sinusoidal due to slew-induced distortion.

Theoretically, $V_{out} = A_v (V_+ - V_-)$, where A_v is the *open-loop voltage gain*. The **closed-loop** gain is a function of the feedback resistor and the input resistor. For the circuit in the previous question (non-saturated state), what gain did you observe? How does this relate to the resistors used in the circuit?

What maximum output swing did you observe? What is this as a function of the rails?

Since you know the equation (6.1) and you know the supply voltages you used, you can calculate the point at which saturation would occur, i.e. for what value of V_{in}? Show the calculation.

Measure the voltage at the inverting input. Sketch the circuit and indicate how you are doing this measurement.

In Figure 6.7, the inverting input is said to be at virtual ground. **A virtual ground** is a voltage ground because the point is at 0 V, however, it is not a current ground because it cannot sink any current.

3. Change the supply voltage to $+5$ V/-12 V. Sketch your circuit and show both the input and output voltages. Do you understand why the supply voltages are called the rails? Note the relationship of V_{out} to the supply voltages; does V_{out} reach the rails?

4. Change the supply voltage to $+7.2$ V/0 V. (What is this configuration called?) Sketch your circuit and show both the input and output voltages. Is this consistent with what happened in the previous item? **What does this illustrate about using inverting amplifiers in single supply configurations?**

5. Return to a supply voltage of ± 12 V. Apply sine waves of different frequencies at input V_{in}. Measure the amplitudes both input and output, calculate the voltage gain at different frequencies. In order to plot a frequency response of the amplifier, you need to fill measurement data in a table like Table 6.2.

Table 6.2   Frequency Response of the Inverting Amplifier
表 6.2   反相放大器的频率响应

Frequency (Hz)	50	100	200	500	1 k	2 k	⋯	500 k	⋯
Input (V)	1	1	1	1	⋯	⋯	⋯	⋯	⋯
Output (V)									
Gain (dB)									

6. Replace R_f with an 1 kΩ resistor, repeat last step. Compare frequency response at different voltage gain.

What happens if you try different resistor values? Hold either the feedback R_f or input resistor R_{in} constant and change the other one. What is the relationship between the resistors and the gain? Do you see the voltage divider formed by R_{in} and R_f?

7. Measure the input and output resistance of the amplifier. To perform these measurements, you need to choice proper test resistors (input series resistor and load) for this circuit.

Test resistor $R_i = $ _____, input resistance $r_i = $ _____.

Load resistor $R_L = $ _____, output resistance $r_o = $ _____.

6.4.4 Non-inverting Amplifier (同相放大器)

The **non-inverting amplifier** circuit (Figure 6.8) produces an output signal V_{out}

that is given by

$$V_{out} = \left(\frac{R_f}{R_{in}} + 1\right) \cdot V_{in}$$

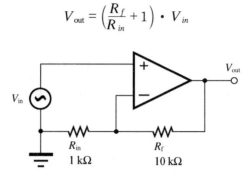

Figure 6.8   A Non-inverting Amplifier
图 6.8   同相放大器

1. Since no signal enters the inverting input terminal, the output signal will be of the **same** polarity as the input signal. If we would like the above circuit to have a gain of ten, what are the relationships between R_{in} and R_f?

   What is the smallest gain which can be produced by this amplifier?

   When R_{in} is infinite, and R_f equals zero, what does the circuit become?

2. Construct the non-inverting amplifier circuit shown in Figure 6.8, subject to the following constraints:

   • the voltage gain is 10

   • select resistors in the 1 kΩ to 100 kΩ range (record the actual resistor values for use in calculations)

   • use ± 12 V for the supply voltages

   • let V_{in} be 100 mV DC

   Measure V_{out}. Considering the input voltage, does the circuit work as you expect? Compare the measured V_{out} with the calculated one.

3. Drive V_{in} with an 1 kHz, 1 V_{PP} sine wave (verify if signal is correct before applying to the circuit). What is the observed maximum output swing? What happens if you choose a 5 V_{PP} signal?

4. Change the supply voltages to + 7.2 V/0 V (single supply configuration), adjust the amplitude and/or DC offset of the input in order to answer the question: What are the limits of the input voltage which produces an undistorted output? Is this consistent with the changes you made in the supply voltages?

   This gives a reasonable limit of the load which can be driven by the op-amp.

6.4.5   Differential Amplifier (差动放大器)

The **differential amplifier** circuit or **difference amplifier** circuit shown in Figure

6.9 (not to be confused with the **differentiator** circuit) produces an output voltage equal to the difference between the two input voltages.

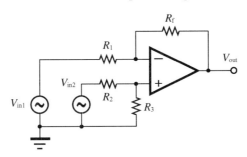

Figure 6.9   A Differential Amplifier Circuit
图 6.9   差动放大器

Note the two voltage dividers in the circuit. In this circuit, typically, $R_1 = R_2$ and $R_3 = R_f$, so that the gain for the inverting and non-inverting inputs is the same. In the simplest case, $V_{out} = V_{in2} - V_{in1}$, when the gain is one.

Construct a summing amplifier circuit on the breadboard using op-amp. Use ± 12 V for the supply voltages. Choose resistors in the range of 1 kΩ to 100 kΩ to obtain a gain of 1. Drive V_{in1} with a time varying signal (for instance, a sine wave) and V_{in2} with a constant voltage. Note the relationship between V_{out} and the inputs. What is the observed maximum output swing?

Change the circuit to get a non-unity gain. What is the observed maximum output swing?

[实验简介]

运算放大器是一个高增益的直流放大器,最早应用于模拟计算机的数学运算.它是一个模拟器件,被广泛运用于模拟和数字电路的设计中.本实验的目的是了解:

1. 运算放大器的符号和术语;
2. 开环结构运算放大器——(比较器)的特性;
3. 闭环结构运算放大器——反相放大器、同相放大器、差动放大器的特性.

6.5 Questions(问题)

1. The voltage gain of a voltage follower is unity ($A_v = 1$). What possible use is an amplifier that doesn't even amplify the voltage of its input signal? Will this circuit be simply replaced by a straight piece of wire?

2. When the two input terminals of an op-amp are shorted together and are connected directly to ground, what should this op-amp's output voltage be? Try this in your lab.

In reality，the behavior of an op-amp is not always the same as what would be ideally predicted. Identify the fundamental problem in real op-amps in this case，and also try to find a solution.

〔Vocabulary〕

operational amplifier（op-amp）：运算放大器

inverting amplifier：反相放大器

non-inverting amplifier：同相放大器

bench supply：桌上电源

comparator：比较器

closed-loop configuration：闭环结构

virtual ground：虚地

non-zero output：非零输出

offset voltage：偏置电压

dual supply：（带正负极性的）双电源供电

single supply：单电源供电

open-loop configuration：开环结构

closed-loop op-amp：闭环运算放大器

time varying signal：时变信号

slew rate：电压转换速率，压摆率

〔Translations〕

1. If it is necessary to compensate for the offset voltage，short the inverting and non-inverting inputs together to ground and adjust the variable resistor until the output is zero.

   如果有必要对漂移电压进行补偿,请将反相输入端和同相输入端短路到地,调节可变电阻,直到输出为零.

2. It is important to note that the supply voltages are indicated as positive and negative. The positive voltage supply must be greater than the negative voltage supply. Any op-amp can be used in either single supply or dual supply configuration，but there is a difference in how each will perform.

   请注意,供电端标有正极和负极,这很重要.正极电压必须高于负极电压.所有运放都可以采用单电源供电也可以采用双电源供电,但二者的性能存在差别.

3. The term open-loop comes from control system theory. There is no feedback in the open-loop mode of a control system. Except comparator，operational amplifiers are usually used in closed-loop mode where there is feedback from the output back to the input.

术语"开环"来自控制理论,在控制系统中,"开环"模式没有反馈.除比较器以外,运算放大器通常运用于闭环方式,此时存在由输出到输入的反馈.

4. In addition,the op-amp IC has two inputs for power. These inputs are typically referred to as supply voltages or the rails. Typically,wiring diagrams will show the supply voltages and schematics will not show the supply voltages. The supply voltages are always required when wiring up an op-amp even if not shown on the schematic.

另外,集成运放还有两个提供电力的输入端,它们一般被称为电源或电源线.通常,连线图上会画出电源,而示意图不会画出.连线运放时,供电电压总是需要的,即使在示意图上没有显示出来.

5. A very important parameter of operational amplifier performance is slew rate. Slew rate is defined as the maximum rate of output voltage change over time $(\frac{\mathrm{d}v}{\mathrm{d}t})$ that an op-amp can output and is given units of $V/\mu s$. Slew rate is measured by applying a large signal step,such as $1\,V$,to the input of the op-amp,and measuring the rate of change from 10% to 90% of the output signal's amplitude. Amplifiers with higher slew rate tend to have higher bandwidth.

运算放大器性能的一个重要参数是压摆率.压摆率定义为运放能够输出的电压相对时间的最大变化速率$(\frac{\mathrm{d}v}{\mathrm{d}t})$,单位是 $V/\mu s$.压摆率的测量方法是,输入一个大的跃变信号,例如 $1\,V$,测量输出信号幅度从 10% 到 90% 的变化速率.较大压摆率的放大器通常有较大的带宽.

6. Since no signal enters the inverting input terminal,the output signal will be of the same polarity as the input signal. If we would like the above circuit to have a gain of ten,what are the relationships between R_{in} and R_{f}? What is the smallest gain which can be produced by this amplifier? When R_{in} is infinite,and R_{f} equals zero,what circuit does this become?

由于没有信号进入反相输入端,输出信号和输入信号将会有相同的极性.如果我们要让上面的电路有 10 倍的增益,那么 R_{in} 和 R_{f} 之间应有什么关系? 该放大器产生的最小增益是多少? 当 R_{in} 为无限大,而 R_{f} 为零时该电路会变成怎样?

7

Differentiators and Integrators
微分器与积分器

Besides voltage gain, any system in a large class known as linear, time-invariant (LTI) is completely characterized by its impulse response. That is, for any input, the output can be calculated in terms of the input and the impulse response.

7.1 Differentiator Amplifier (微分器)

7.1.1 Op-amp Differentiator Circuit (运放构成的微分器电路)

Figure 7.1 shows a operational amplifier differentiator. As its name implies, the basic operational amplifier differentiator circuit produces an output signal which is proportional to the input voltage's rate-of-change with respect to time, therefore performs the mathematical operation of differentiation.

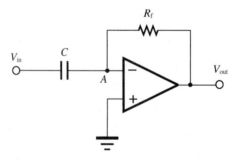

Figure 7.1 Op-amp Differentiator Circuit

图 7.1 运放构成的微分器电路

The input signal to the differentiator is applied to the capacitor. The capacitor blocks any DC content so there is no current flow to the amplifier summing point (node A), resulting in zero output voltage. The capacitor only allows AC type input voltage changes to pass through and whose frequency is dependant on the rate of change of the input signal.

Since the node voltage of the operational amplifier at its inverting input terminal is zero (virtual ground), the current i flowing through the capacitor will be given as

$$i = -\frac{V_{out}}{R_f} \tag{7.1}$$

The charge on the capacitor equals capacitance times voltage across the capacitor, i.e. $Q = C \cdot V_{in}$. Thus the rate of change of this charge is:

$$\frac{dQ}{dt} = C\frac{dV_{in}}{dt} \tag{7.2}$$

but dQ/dt is the capacitor current i, thus

$$-\frac{V_{out}}{R_f} = C\frac{dV_{in}}{dt} \tag{7.3}$$

from which we have an ideal voltage output for the op-amp differentiator is given as

$$V_{out} = -R_f C\frac{dV_{in}}{dt} \tag{7.4}$$

7.1.2 Impulse Response (冲激响应)

If we apply a constantly changing signal such as a square wave, triangular wave or sine wave signal to the input of a differentiator amplifier circuit, the resultant output signal will be changed and whose final shape is dependant upon the RC time constant of the resistor/capacitor combination (Figure 7.2).

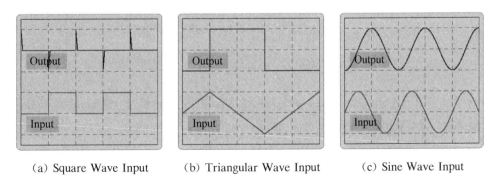

(a) Square Wave Input   (b) Triangular Wave Input   (c) Sine Wave Input

Figure 7.2　Differentiator Output at Different Input Signal
图 7.2　不同输入信号的微分器输出

To generally describe the reaction of any dynamic system in response to some external change, **impulse response** is introduced. The impulse response of a dynamic system $h(t)$ is its output when presented with a brief input signal, called an impulse. The impulse can be modeled as a **Dirac delta function** for continuous-time systems. The Dirac delta $\delta(t)$ can be loosely thought of as a function on the real line which is zero everywhere except at $t = 0$, where it is infinite,

$$\delta(t) = \begin{cases} +\infty, & t = 0 \\ 0, & t \neq 0 \end{cases} \tag{7.5}$$

and which is also constrained to satisfy the identity

$$\int_{-\infty}^{\infty} \delta(t)\mathrm{d}t = 1 \tag{7.6}$$

In Fourier analysis, such an impulse comprises equal portions of all possible excitation frequencies, which makes it a convenient test probe. The impulse response of a circuit, in the time domain, is proportional to the bandwidth of the circuit in the frequency domain.

7.1.3 Step Response (阶跃响应)

The impulse function is defined as an infinitely high, infinitely narrow pulse, with an area of unity. This is, of course, impossible to realize in a physical sense. The **step response** provides a convenient way to figure out the impulse response of a system.

The unit-step input is defined as:

$$u(t) = \begin{cases} 0, & t < 0 \\ 1, & t \geqslant 0 \end{cases} \tag{7.7}$$

The step response of a system is defined as its response to a unit-step input $u(t)$. Knowing impulse response $h(t)$, the step response $g(t)$ can be obtained by integrating the impulse response:

$$g(t) = \int_{0}^{t} h(t - \tau)\mathrm{d}\tau \tag{7.8}$$

Compared to the Dirac impulse, step impulse is much easier to practical use, as the rising edge of a square wave can be used as an input test signal.

7.1.4 Frequency Response (频率响应)

In case of steady-state of Figure 7.1, the reactance of a capacitor C is $X_C = 1/\mathrm{j}\omega C$. The current flowing through the capacitor is given as

$$i = V_{\mathrm{in}} \cdot \mathrm{j}\omega C \tag{7.9}$$

Combining equation (7.9) with equation (7.1) gives

$$A(\mathrm{j}\omega) = \frac{V_{\text{out}}}{V_{\text{in}}} = -\mathrm{j}\omega R_{\text{f}} C \qquad (7.10)$$

$A(\mathrm{j}\omega)$ is **frequency response** of the differentiator.

Both impulse response and frequency response characterize a **linear time-invariant system**. More generally, the impulse response describes the reaction of the system as a function of time, whereas the frequency response describes the reaction of the system as a function of frequency.

7.1.5 Practical Differentiator Amplifier (实际微分放大器)

The basic single resistor and single capacitor op-amp differentiator circuit shown in Figure 7.1 is not widely used to reform the mathematical function of differentiation. The circuit has limitations at high frequencies:

- unstable and easy to oscillate;
- sensitive to noise.

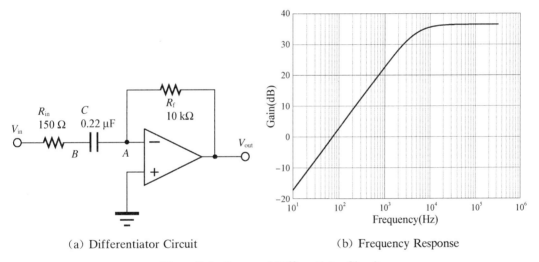

(a) Differentiator Circuit　　　　(b) Frequency Response

Figure 7.3　Improved Differentiator Circuit

图 7.3　改进的微分器电路

In order to reduce the overall closed-loop gain of the circuit at high frequencies, an extra resistor R_{in} is connected in series with the capacitor C, which limits the increase in gain to a ratio of $R_{\text{f}}/R_{\text{in}}$ at high frequencies (Figure 7.3). The circuit acts like a differentiator at low frequencies, as the reactance of the capacitor is much greater than that of the resistor R_{in}, and an amplifier with resistive feedback at high frequencies where the capacitor's resistance is negligible, giving much better noise rejection.

Since negative feedback is present through the resistor R_{f}, we can apply the virtual ground concept to analyze the circuit.

$$\frac{V_{\text{in}}}{R_{\text{in}} + \dfrac{1}{j\omega C}} = -\frac{V_{\text{out}}}{R_{\text{f}}} \tag{7.11}$$

therefore

$$A(j\omega) = \frac{V_{\text{out}}}{V_{\text{in}}} = -\frac{j\omega R_{\text{f}} C}{1 + j\omega R_{\text{in}} C} \tag{7.12}$$

The **frequency response** according to equation (7.12) is shown in Figure 7.3(b). The circuit is also used as the first order active **high pass filter**.

7.1.6 Simulation (仿真)

Listing 7.1 is the netlist of a differentiator (Figure 7.3) using OP07. "op07. mod" is the script file for OP07 (element "X1") which can be found in your NGSPICE software package. The model file needs to be put in current working directory. The op-amp's model defines nodes as V_{in}^{+}, V_{in}^{-}, V_{CC}, V_{EE} and V_{out} in order. The simulation results are shown in Figure 7.4.

Listing 7.1 Netlist of Op-amp Differentiator
清单 7.1 运放微分器的网连表单

```
1    OP-amp Differentiator Using OP07
2    .include op07.mod
3
4    Vin in 0 dc 0 ac 1 PULSE(0V 0.2V 100us 5us 5us 1ms 2ms)
5    Rin in B 150
6    C B A 0.22u
7    R A out 10k
8    X1 0 A VCC VEE out OP07
9    Vcc VCC 0 12V
10   Vee VEE 0 - 12V
11
12   .control
13   set color0= white
14   set color1= black
15   tran 1us 2ms
16   run
17   plot v(out) {20* v(in)}
18   ac dec 10 10 1Meg
19   plot db(v(out))
20   .endc
21   .end
```

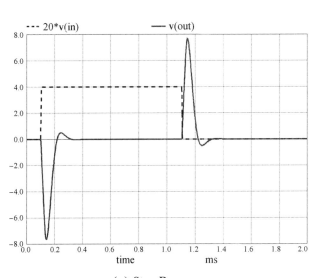

(a) Step Response

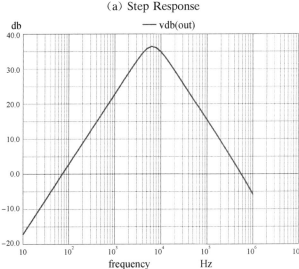

(b) Frequency Response

Figure 7.4　Transient Analysis and Frequency Response of Differentiator

图 7.4　微分器的瞬态分析和频率响应

7.2　Operational Amplifier Integrator（运放构成的积分器）

An op-amp integrator is an operational amplifier circuit that performs the mathematical operation of integration，that is we can cause the output to respond to changes in the input voltage over time as the op-amp integrator produces an output voltage which is proportional to the integral of the input voltage.

7.2.1　Mathematical Analysis of an Ideal Integrator（理想积分器的数学分析）

As we can see in Figure 7.5，the resistor R is connected to the input terminal of the inverting amplifier while the capacitor C forms the negative feedback element across the operational amplifier. When a step voltage, V_{in} is firstly applied to the input of an integrating amplifier, the uncharged capacitor C has very little resistance and acts a bit like a short circuit allowing maximum current to flow via the input resistor R. No current flows into the amplifiers input and the inverting input is a virtual ground resulting in zero output.

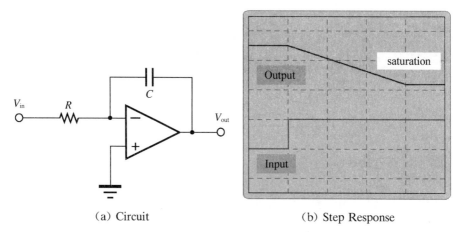

(a) Circuit   (b) Step Response

Figure 7.5   Op-amp Integrator

图 7.5   运放构成的积分器

As the impedance of the capacitor at this point is very low, the gain ratio of X_C/R is also very small, giving an overall voltage gain of less than one (voltage follower circuit). Due to the influence of the input voltage, the feedback capacitor C begins to charge up, its impedance slowly increase in proportion to its rate of charge. Since the capacitor is connected between the op-amp's inverting input (which is virtual grounded) and the op-amp's output (which is now negative), the potential voltage, V_C developed across the capacitor slowly increases causing the charging current to decrease as the impedance of the capacitor increases.

Recall that the voltage across a capacitor is equal to the charge on the capacitor divided by its capacitance giving $V_C = Q/C$. The charge rate is proportional to the voltage changing across the capacitor:

$$\frac{\mathrm{d}Q}{\mathrm{d}t} = -C\frac{\mathrm{d}V_{out}}{\mathrm{d}t} \tag{7.13}$$

The capacitor current $\mathrm{d}Q/\mathrm{d}t$ is also the current flowing through the resistor R, gives

$$\frac{V_{in}}{R} = -C\frac{\mathrm{d}V_{out}}{\mathrm{d}t} \tag{7.14}$$

From which we derive an ideal voltage output for the op-amp integrator as

$$V_{out} = -\frac{1}{RC}\int V_{in}\mathrm{d}t \tag{7.15}$$

The output voltage is a constant $1/RC$ times the integral of the input voltage with respect to time.

As input voltage unchanged after rising edge, the capacitor gradually becomes fully charged and acts as an open circuit, blocking any more flow of DC current. The reactance of the capacitor is infinite resulting in infinite gain. The result is that the output of the amplifier goes into saturation [Figure 7.5(b)].

The rate at which the output voltage increases is determined by the value of the resistor and the capacitor, "RC time constant". By changing this RC time constant value, either by changing the value of the capacitor C or the resistor R, the time in which it takes the output voltage to reach saturation can also be changed.

If we apply a constantly changing input signal such as a square wave to the input of an integrator amplifier, the capacitor will charge and discharge in response to changes in the input signal. This results in the output signal being that of a triangular waveform whose output is affected by the RC time constant of the resistor/capacitor combination because at higher frequencies, the capacitor has less time to fully charge. This type of circuit is also known as a **ramp generator**.

7.2.2 Frequency Response of the Integrator (积分器的频率响应)

If we use a sine wave as the input signal, the op-amp integrator performs less like an integrator and begins to behave more like an active low pass filter, passing low frequency signals while attenuating the high frequencies.

Assume $V_{in} = \mathrm{e}^{j\omega t}$, from equation (7.15), gives

$$V_{out} = -\frac{1}{j\omega RC}\mathrm{e}^{j\omega t} \tag{7.16}$$

thus

$$A(j\omega) = \frac{V_{out}}{V_{in}} = -\frac{1}{j\omega RC} \tag{7.17}$$

We can see from equation (7.17) that the gain $|A(j\omega)|$ is inversely proportional to frequency ω. Therefore, with just a single capacitor C in the feedback path, at zero frequency the op-amp is effectively connected as a normal open-loop amplifier with very high open-loop gain. This results in the op-amp becoming unstable cause undesirable output voltage conditions and possible voltage rail saturation.

7.2.3 Practical Integrator (实用型积分器)

In Figure 7.6(a), a high value resistance R_2 in parallel with capacitor is connected in the circuit. The result is, at normal operating frequencies the circuit acts as a standard integrator, while at very low frequencies approaching, when C becomes open-circuited due to its reactance $1/j\omega C$, the magnitude of the voltage gain is limited and controlled by the ratio of R_2/R_1.

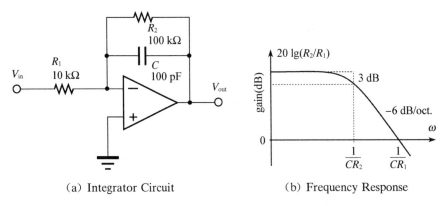

(a) Integrator Circuit　　　　　(b) Frequency Response

Figure 7.6　The AC Integrator with DC Gain Control

图 7.6　带有直流增益控制的交流积分器

The output voltage of the integrator amplifier at any instant will be the integral of an input waveform, so that when the input is a square wave, the output waveform will be triangular. A sinusoidal input waveform will produce another sine wave as its output which will be $90°$ out-of-phase with the input producing a cosine wave.

Further more, when the input is triangular near the **corner frequency**, the output waveform may also be sinusoidal. This then forms the basis of an active **low pass filter**, as the harmonics in high-frequency are attenuated by the amplifier.

7.3　Objectives（实验目的）

The objectives for this experiment are:

1. to study the op-amp differentiator.
   - building an op-amp differentiator.
   - investigating the step response of the amplifier.
   - investigating the frequency response of a practical differentiator.
2. to study the op-amp integrator:
   - building an op-amp integrator.
   - investigating the output voltage waveform with input waveform.
3. to understand the characteristics of a system both in time domain and in frequency domain.

7.4 Experimental Procedure (实验步骤)

7.4.1 Differentiator(微分器)

Study the op-amp differentiator according to Figure 7.3(a), perform following measurement:

1. Short-circuit R_{in} temporarily, set voltage supply of ± 12 V to the op-amp. Apply square wave of 100 mV$_{PP}$/100 Hz to the input, record the output waveform using oscilloscope.

2. Apply triangular wave of 100 mV$_{PP}$/100 Hz to the input, watch the output waveform and sketch it in your notebook. Note down the amplitude of the waveform.

   What should the output amplitude be, according to your theoretical analysis?

3. Watch the output while increasing the frequency of input signal, read the output magnitude. At what frequency does the output saturate? Give your analysis.

4. Add R_{in} to the circuit, repeat the measurements above.

5. Apply sinusoidal input instead of triangular wave, measure the frequency response of the circuit, plot the amplitude response of the differentiator. You may need a table like Table 6.2 to collect your data.

   Find the corner frequency (-3 dB point) on the frequency response curve:
   $f_c = $ _____
   The circuit is also used as the first order high pass filter with cutoff frequency f_c.

7.4.2 Integrator(积分器)

Build and investigate op-amp integrator as in Figure 7.6(a).

1. Open R_2 temporarily, set voltage supply of ± 12 V to the op-amp. Apply square wave of 100 mV$_{PP}$/10 kHz to the input, watch the output waveform using oscilloscope. Note down the amplitude and frequency of the waveform.

   What is the amplitude of the output waveform, according to theoretical analysis?

2. Watch the output while decreasing the frequency of input signal, read the output magnitude. At what frequency does the output saturate? Give your analysis.

3. Add R_2 to the circuit，apply sinusoidal input instead of square wave，measure the frequency response of the circuit，plot the amplitude response of the integrator.

Find the corner frequency (－3 dB point) on the frequency response curve of the first order low pass filter：

$f_c =$ _____

［实验简介］

一个电子系统包含时间和频率两个方面的特性. 描述一个系统时域特性的是冲激响应函数 $h(t)$，频率响应 $H(j\omega)$ 描述该系统的频率特性，二者通过傅里叶变换建立联系.

微分器和积分器是典型的波形变换电路，同时也是典型的一阶高通/低通滤波器. 通过对这两个电路的研究，理解一个系统在时间域和频率域的表现. 由此开始，建立线性非时变系统的概念.

［Vocabulary］

differentiator：微分器

linear，time-invariant system：线性非时变（缩写为 LTI）系统

impulse response：冲激响应

block：阻挡

RC time constant：*RC* 时间常数

step response：阶跃响应

high pass filter：高通滤波器

Dirac delta function：狄拉克 δ 函数

reactance：电抗（于电容是容抗，于电感是感抗，与电阻单位相同）

integrator：积分器

low pass filter：低通滤波器

time domain：时域

frequency domain：频域

狄拉克 δ 函数，简称 δ 函数，数学上被定义为时间无穷短、幅度无穷大、具有有限能量的瞬时脉冲信号. 它的频谱是一条水平直线，即等幅度地包含了所有的频率成分. 测量一个系统对 δ 函数的响应，通过傅里叶变换即可获得该系统的频率特性. 但是严格的 δ 函数在物理上是不可实现的，产生阶跃信号则相对比较容易，且冲激响应函数 $h(t)$ 可以通过阶跃响应 $a(t)$ 的微分得到，即

$$h(t) = \frac{\mathrm{d}}{\mathrm{d}t}a(t)$$

[Translations]

1. In Fourier analysis，such an impulse comprises equal portions of all possible excitation frequencies，which makes it a convenient test probe. The impulse response of a circuit，in the time domain，is proportional to the bandwidth of the circuit in the frequency domain.

   在傅里叶分析中,这样的脉冲均匀包含了所有的激励频率成分,这使得它成为一个方便的测试(工具). 一个电路在时域的冲激响应与其在频率上的带宽正相关.

2. The impulse function is defined as an infinitely high，infinitely narrow pulse，with an area of unity. This is，of course，impossible to realize in a physical sense.

   冲激函数被定义为无限高、无限窄的单位面积脉冲,这当然在物理上是不能实现的.

3. Both impulse response and frequency response characterize a linear time-invariant system. More generally， the impulse response describes the reaction of the system as a function of time，whereas the frequency response describes the reaction of the system as a function of frequency.

   冲激响应和频率响应二者都描述了一个线性非时变系统的特征. 一般来说,冲激响应描述了系统关于时间的响应函数,而频率响应描述了系统关于频率的响应函数.

4. In order to reduce the overall closed-loop gain of the circuit at high frequencies，an extra resistor R_{in} is connected in series with the capacitor C，which limits the increase in gain to a ratio of R_f/R_{in} at high frequencies.

   为了降低电路在高频段的整体闭环增益,接入了一个与电容 C 串联的电阻 R_{in},用于将其在高频段的增益限 制在 R_f/R_{in} 比率之内.

5. The output voltage of the integrator amplifier at any instant will be the integral of an input waveform，so that when the input is a square wave，the output waveform will be triangular. A sinusoidal input waveform will produce another sine wave as its output which will be 90° out-of-phase with the input producing a cosine wave.

   在任何时刻,积分放大器的输出电压是输入波形的积分,因此当输入为方波时,输出波形是三角波. 正弦输入波形将产生另一个正弦波(输出),其相位与输入相差 90°,也就是余弦波形.

8

Active Filters
有源滤波器

A filter is a network that processes signals in a frequency-dependent manner. Filters can be used to separate signals, passing those of interest, and attenuating the unwanted frequencies.

In this lab, you will be investigating some of active filters, these include:

- Low pass filters: attenuation of frequencies above their cutoff points, only frequencies lower than cutoff frequency may pass through the system.
- High pass filters: attenuation of frequencies below their cutoff points, only frequencies higher than cutoff frequency are allowed to pass through the system.
- Bandpass filters: attenuation of frequencies both above and below those they allow to pass, frequencies in a given range may pass through the system.
- Bandstop filters or band-rejection filters: attenuation of frequencies only in a specific range, most frequencies other than stopband are allowed to pass through the system.

8.1 Introduction (简介)

Filter circuits are used in a wide variety of applications. A simple, single pole, high pass filter (the differentiator) can be used to block DC offset in high gain amplifiers or single supply circuits. A simple, single pole, low pass filter (the integrator) is often used to stabilize amplifiers by rolling off the gain at higher frequencies where excessive phase shift may cause oscillations.

In the field of telecommunication, bandpass filters are used in the audio frequency range (20 Hz to 20 kHz) for modems and speech processing. High frequency bandpass filters (several hundred MHz) are used for channel selection in telephone central offices. Data acquisition systems usually require anti-aliasing low pass filters as well as low pass noise filters in their preceding signal conditioning stages. Power supplies of a system often use band-rejection filters to suppress the 50 Hz/60 Hz **power frequency** and high frequency transients.

In addition，there are filters that do not filter any frequencies of a complex input signal，but just add a linear phase shift to each frequency component，thus contributing to a constant time delay. These are called all-pass filters.

8.1.1 Frequency Response of Idealized Filters（理想滤波器的频率响应）

An ideal filter will have a fixed gain of amplitude response for the frequencies of interest and zero everywhere else. The frequency range that is allowed to pass is called passband，whereas the range that signal is greatly attenuated is called stopband. The frequency at which the response changes from passband to stopband is referred to as the **cutoff frequency**. Figure 8.1 shows the frequency response of some idealized filter types.

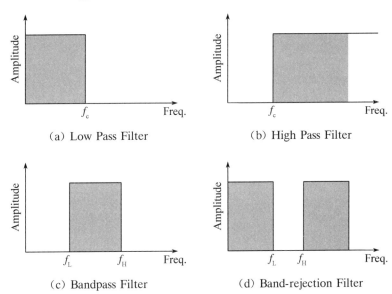

(a) Low Pass Filter　　　　　　　　(b) High Pass Filter

(c) Bandpass Filter　　　　　　　(d) Band-rejection Filter

Figure 8.1　Frequency Response of Idealized Filters

图 8.1　理想滤波器的频率响应

8.1.2 Realistic Filters（可实现滤波器）

Unfortunately，the idealized filters cannot be easily built because of **causality constraints**. The transition from passband to stopband will not be instantaneous，instead，there will be a transition region. Also，the amplitude response in stopband will not be zero. Therefore，in a practical filter designing，you have to make a compromise between **transition bandwidth**, **passband ripple** and **stopband attenuation** (Figure 8.2).

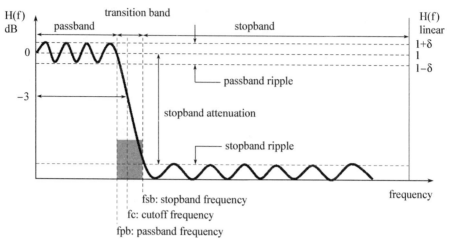

Figure 8.2   Realistic Filter Design

图 8.2   实际滤波器设计

The steepness of a filter is affected by **the order of the filter**. Order is also the number of poles in the transfer function. A pole is a root of the denominator of the transfer function. A pole frequency corresponds to a corner frequency at which the slope of the magnitude curve decreases by -6 dB/octave, or -20 dB/decade. Conversely, a zero is a root of the numerator of the transfer function. A zero corresponds to a corner frequency in the magnitude response at which the slope increases by 6 dB/octave, or 20 dB/ decade. Pole-zero combination forms the passband and stopband of a filter.

Although the amplitude response of a filter is discussed in designing, it should be noticed that the filter will also affect the phase of a signal, as well as the amplitude. A phase shifter is a type of filter that is designed only for changing the phase.

8.1.3   Active Filters(有源滤波器)

Filters only consist of passive components such as inductors, resistors and capacitors are called passive filters or LRC filters [Figure 8.3(a)]. In the lower frequency range (below 1 MHz), however, the inductor value becomes very large and the inductor itself gets quite bulky, making economical production difficult. In these cases, active filters become important. Active filters are circuits that use an operational amplifier as the active device in combination with some resistors and capacitors to provide an LRC-like filter performance [Figure 8.3(b)].

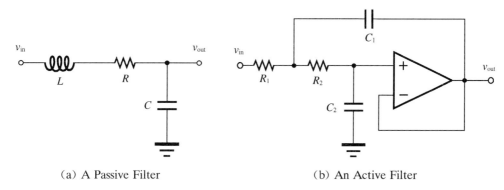

(a) A Passive Filter   (b) An Active Filter

Figure 8.3   Passive and Active Filter

图 8.3   无源滤波器和有源滤波器

8.2   Filter Designing (滤波器设计)

8.2.1   Sallen-Key Topology (萨兰-基结构)

The Sallen-Key topology (named after R. P. Sallen and E. L. Key of MIT's Lincoln Labs) is an electronic filter topology used to implement second-order active filters that is particularly valued for its simplicity. It is a degenerate form of a **voltage-controlled voltage-source** (VCVS) filter topology. Sallen-Key filter is a variation on a VCVS filter that uses a unity-voltage-gain amplifier (Figure 8.4). Filters with higher order can be obtained by cascading two or more sections.

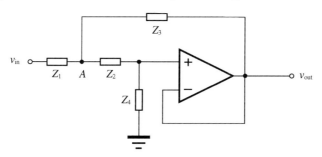

Figure 8.4   The Generic Sallen-Key Topology

图 8.4   萨兰-基拓扑结构

The following analysis to Figure 8.4 is based on the assumption that the operational amplifier is ideal.

Because the op-amp is in a negative-feedback configuration, its v_+ and v_- inputs must be the same (virtual short), i. e. $v_+ = v_-$. However, the inverting input v_- is connected directly to the output v_{out}, we have

▶▶▶

$$v_+ = v_- = v_{out} \tag{8.1}$$

Applying Kirchhoff's current law at node A,

$$\frac{v_{in} - v_A}{Z_1} = \frac{v_A - v_{out}}{Z_3} + \frac{v_A - v_+}{Z_2} \tag{8.2}$$

Combining equation (8.1) and equation (8.2),

$$\frac{v_{in} - v_A}{Z_1} = \frac{v_A - v_{out}}{Z_3} + \frac{v_A - v_{out}}{Z_2} \tag{8.3}$$

Applying equation (8.1) and Kirchhoff's current law at the op-amp's non-inverting input v_+ gives

$$\frac{v_A - v_{out}}{Z_2} = \frac{v_{out}}{Z_4} \tag{8.4}$$

which means that

$$v_A = v_{out}\left(\frac{Z_2}{Z_4} + 1\right) \tag{8.5}$$

Combining equation (8.2) and equation (8.4) gives

$$\frac{v_{in} - v_{out}\left(\frac{Z_2}{Z_4} + 1\right)}{Z_1} = \frac{v_{out}\left(\frac{Z_2}{Z_4} + 1\right) - v_{out}}{Z_3} + \frac{v_{out}\left(\frac{Z_2}{Z_4} + 1\right) - v_{out}}{Z_2} \tag{8.6}$$

Rearranging equation (8.6) gives the transfer function

$$\frac{v_{out}}{v_{in}} = \frac{Z_3 Z_4}{Z_1 Z_2 + Z_3(Z_1 + Z_2) + Z_3 Z_4} \tag{8.7}$$

By choosing different passive components (e. g. resistors and capacitors) for Z_1, Z_2, Z_3 and Z_4, the filter can be made with low pass, bandpass, and high pass characteristics. Recall that a capacitor with capacitance C has impedance $Z_C = 1/sC$, where $s = \sigma + j\omega$ is the complex frequency variable. For pure sine waves, the damping constant σ becomes zero and $s = j\omega = j2\pi f$.

8.2.2 Low Pass Filter (低通滤波器)

The generic transfer function of a low pass filter can be written as

$$A(s) = \frac{A_0 \omega_L^2}{s^2 + \frac{\omega_L}{Q}s + \omega_L^2} \tag{8.8}$$

where A_0 is the gain at low frequencies, ω_L is the angular eigenfrequency (also called cutoff frequency by low pass or high pass applications), Q is the quality factor (or Q-factor). The Q-factor determines the height and width of the peak of the frequency response of the filter (Figure 8.5). For band-pass filters, Q is defined as the ratio of the central frequency f_c to the bandwidth at the two $-3\,\mathrm{dB}$ points.

$$Q = \frac{f_c}{f_H - f_L}.$$

For example, a Butterworth[1] filter (also known as the maximally flat magnitude filter), which has maximally flat passband frequency response, has a Q-factor of $1/\sqrt{2}$. By comparison, a value of $Q = 1/2$ corresponds to the series cascade of two identical simple low pass filters.

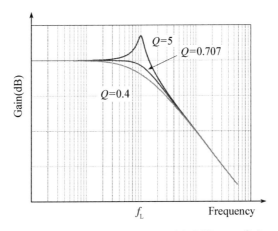

Figure 8.5 Frequency Response with Different Q-factor

图 8.5 不同 Q 值的频率响应

From equation (8.8), the magnitude of the gain response is

$$|A(j\omega)| = \frac{A_0}{\sqrt{\left(\dfrac{\omega^2}{\omega_L^2} - 1\right)^2 + \left(\dfrac{\omega}{Q\omega_L}\right)^2}} \qquad (8.9)$$

For angular frequencies $\omega \gg \omega_L$, $|A(j\omega)| \approx \dfrac{\omega_L^2}{\omega^2}$, i.e. the rolloff of the second-order low pass filter is 12 dB/octave, or 40 dB/decade. In general, n-th order low pass filter gives $6n$ dB/octave of rolloff.

An example of a non-unity-gain low pass configuration is shown in Figure 8.6. This circuit is equivalent to the generic case above with $Z_1 = R_1$, $Z_2 = R_2$, $Z_3 = 1/sC_1$ and $Z_4 = 1/sC_2$. Notice how R_p and R_4 are arranged in a non-inverting amplifier configuration, gives a gain of $A_0 = R_p/R_4 + 1$. Thus,

---

① Stephen Butterworth (Aug. 11, 1885 – Oct. 28, 1958), British physicist and engineer.

▶ ▶ ▶

$$\begin{cases} A_0 = \dfrac{R_p}{R_4} + 1 \\[2mm] \omega_L = 2\pi f_L = \dfrac{1}{\sqrt{R_1 R_2 C_1 C_2}} \\[2mm] Q = \dfrac{\sqrt{R_1 R_2 C_1 C_2}}{C_2(R_1 + R_2) + R_1 C_1(1 - A_0)} \end{cases} \qquad (8.10)$$

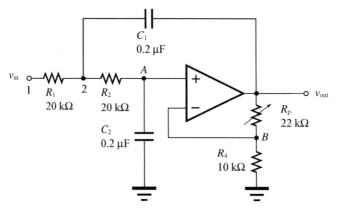

Figure 8.6　Second-order Low Pass Filter

图 8.6　二阶低通滤波器

By choosing $R_1 = R_2 = R$, $C_1 = C_2 = C$, the cutoff frequency and Q-factor of the low pass filter can be expressed as

$$\begin{cases} f_L = \dfrac{1}{2\pi RC} \\[2mm] Q = \dfrac{1}{3 - A_0} \end{cases} \qquad (8.11)$$

The Sallen-Key circuit has the advantage that the Q-factor can be adjusted by changing the inner gain A_0 without modifying the corner frequency ω_L. A drawback is, however, that Q and A_0 cannot be adjusted independently. Since Q is always positive, the amplitude gain A_0 cannot exceed the value of 3. On the other hand, low gain leads to small Q-factor.

Another prototype is the multiple feedback topology which is commonly used in filter designing. This type of filters may have higher amplitude gain and Q-factors. Figure 8.7(a) shows a second-order multiple feedback low pass filter with $|A_0| = 10$ and $Q = 1.03$. The transfer function is given as

$$A(s) = \dfrac{-\dfrac{1}{R_1 R_3 C_1 C_2}}{s^2 + \dfrac{1}{C_2}\left(\dfrac{1}{R_1} + \dfrac{1}{R_2} + \dfrac{1}{R_3}\right)s + \dfrac{1}{R_2 R_3 C_1 C_2}} \qquad (8.12)$$

Figure 8.7(b) shows its amplitude response, Refer to formula (8.8), its parameters are described as follows

$$\begin{cases} A_0 = -\dfrac{R_2}{R_1} \\[2mm] \omega_L = 2\pi f_L = \dfrac{1}{\sqrt{R_2 R_3 C_1 C_2}} \\[2mm] Q = \dfrac{\sqrt{R_2 R_3 C_1 C_2}}{C_1(R_2 + R_3 + R_2 R_3/R_1)} \end{cases} \qquad (8.13)$$

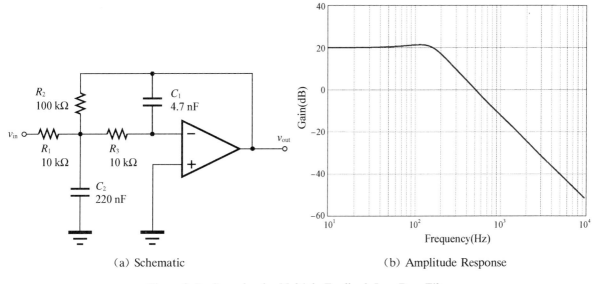

(a) Schematic   (b) Amplitude Response

Figure 8.7   Second-order Multiple Feedback Low Pass Filter

图 8.7　二阶多反馈低通滤波器

8.2.3　High Pass Filter（高通滤波器）

A second-order high pass filter is shown in Figure 8.8.

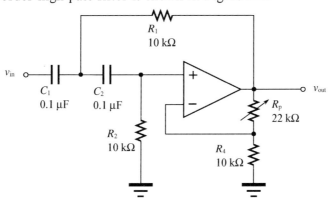

Figure 8.8   Second-order High Pass Filter

图 8.8　二阶高通滤波器

A second-order non-unity-gain high pass filter has the transfer function

$$A(s) = \frac{A_0 s^2}{s^2 + \dfrac{\omega_H}{Q}s + \omega_H^2} \tag{8.14}$$

where cutoff angular frequency ω_H and Q-factor have been discussed in the previous section. Its amplitude response can be represented as

$$|A(j\omega)| = \frac{A_0}{\sqrt{\left(\dfrac{\omega_H^2}{\omega^2} - 1\right)^2 + \left(\dfrac{\omega_H}{Q\omega}\right)^2}} \tag{8.15}$$

The circuit in Figure 8.8 implements this transfer function by the equations

$$\begin{cases} A_0 = \dfrac{R_p}{R_4} + 1 \\[2mm] \omega_H = 2\pi f_H = \dfrac{1}{\sqrt{R_1 R_2 C_1 C_2}} \\[2mm] Q = \dfrac{\sqrt{R_1 R_2 C_1 C_2}}{R_1(C_1 + C_2) + R_2 C_2(1 - A_0)} \end{cases} \tag{8.16}$$

By choosing $R_1 = R_2 = R$, $C_1 = C_2 = C$, the cutoff frequency and Q-factor of the high pass filter can be expressed as

$$\begin{cases} f_H = \dfrac{1}{2\pi RC} \\[2mm] Q = \dfrac{1}{3 - A_0} \end{cases} \tag{8.17}$$

A second-order multiple feedback high pass filter, which can give higher gain and Q-factors, is shown in Figure 8.9(a). The transfer function is given as

$$A(s) = \frac{-\dfrac{C_1}{C_2} s^2}{s^2 + \dfrac{C_1 + C_2 + C_3}{R_1 C_2 C_3}s + \dfrac{1}{R_1 R_2 C_2 C_3}} \tag{8.18}$$

When component values are selected as in the figure, the filter gives Q-factor of 0.8 and corner frequency at 1500 Hz. The corresponding parameters in formula (8.14) are as follows

$$\begin{cases} A_0 = \dfrac{C_1}{C_2} \\[2mm] \omega_H = 2\pi f_H = \dfrac{1}{\sqrt{R_1 R_2 C_2 C_3}} \\[2mm] Q = \dfrac{\sqrt{R_1 R_2 C_2 C_3}}{R_2(C_1 + C_2 + C_3)} \end{cases} \tag{8.19}$$

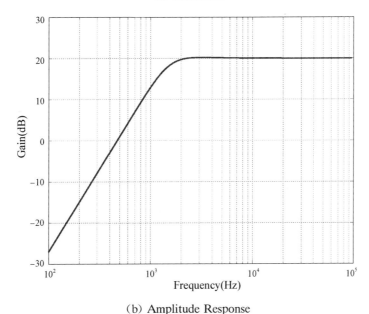

（a）Schematic

（b）Amplitude Response

Figure 8.9　Second-order Multiple Feedback High Pass Filter

图 8.9　二阶多反馈高通滤波器

8.2.4　Bandpass Filter（带通滤波器）

The simplest design of a bandpass filter is to connect a high pass filter and a low pass filter in series，which is commonly done in wide-band filter applications. Thus，a first-order high pass and a first-order low pass provide a second-order bandpass，while a second-order high pass and a second-order low-pass result in a fourth-order bandpass response. As a result，a second-order high pass filter gives the rolloff of 6 dB/octave both at low frequency and at high frequency.

A non-unity-gain bandpass filter implemented with a VCVS filter is shown in Figure 8. 10. Although there is a slight difference comparing with the generic Sallen-Key topology，it can be analyzed using similar methods as above. Its transfer

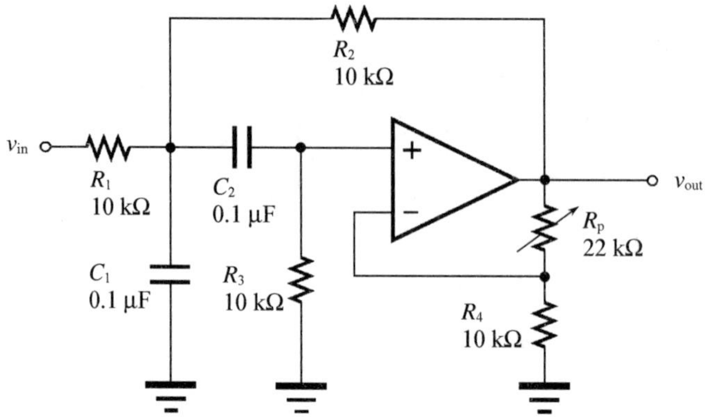

Figure 8.10　Second-order Bandpass Filter

图 8.10　二阶带通滤波器

function is given by

$$A(s) = \frac{A_0 \dfrac{\omega_\text{p}}{Q}s}{s^2 + \dfrac{\omega_\text{p}}{Q}s + \omega_\text{p}^2} \tag{8.20}$$

The amplitude response of the second-order band pass filter is given as

$$|A(\text{j}\omega)| = \frac{A_0}{\sqrt{1 + Q^2 \left(\dfrac{\omega_\text{p}}{\omega} - \dfrac{\omega}{\omega_\text{p}} \right)^2}} \tag{8.21}$$

where

$$\begin{cases} \omega_\text{p} = \sqrt{\dfrac{R_1 + R_2}{R_1 R_2 R_3 C_1 C_2}} \\[2ex] Q = \dfrac{\sqrt{R_1 + R_2}\sqrt{R_1 R_2 R_3 C_1 C_2}}{R_1 R_2 (C_1 + C_2) + R_2 R_3 C_2 + R_1 R_3 C_2 (1 - A_{VF})} \\[2ex] A_0 = \dfrac{R_2 R_3 C_2 A_{VF}}{R_1 R_2 (C_1 + C_2) + R_2 R_3 C_2 + R_1 R_3 C_2 (1 - A_{VF})} \end{cases} \tag{8.22}$$

The voltage gain of the filter is adjustable via A_{VF}, $A_{VF} = 1 + R_\text{p}/R_4$. By choosing $R_1 = R_2 = R_3 = R$, $C_1 = C_2 = C$, the central frequency and Q-factor of the bandpass filter can be expressed as

$$f_\text{p} = \frac{\sqrt{2}}{2\pi RC}$$

$$Q = \frac{\sqrt{2}}{4 - A_{VF}}$$

The amplitude gain at central frequency is $A_0 = \dfrac{A_{VF}}{4 - A_{VF}}$.

A multiple feedback bandpass circuit in Figure 8.11 has the following transfer

────◄◄◄────

function：

$$A(s) = \frac{-R_2 R_3 C_2 s}{R_1 R_2 R_3 C_1 C_2 s^2 + R_1 R_3 (C_1 + C_2) s + (R_1 + R_3)} \quad (8.23)$$

The coefficient comparison with equation (8.20) yields the following equations

$$\begin{cases} A_0 = -\dfrac{C_2}{C_1 + C_2} \dfrac{R_2}{R_1} \\[2mm] \omega_p = 2\pi f_p = \sqrt{\dfrac{R_1 + R_3}{R_1 R_2 R_3 C_1 C_2}} \\[2mm] Q = \dfrac{C_1 C_2}{C_1 + C_2} R_2 \omega_p \end{cases} \quad (8.24)$$

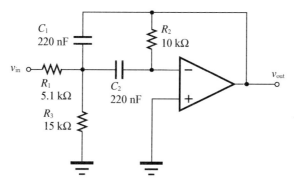

(a) Schematic

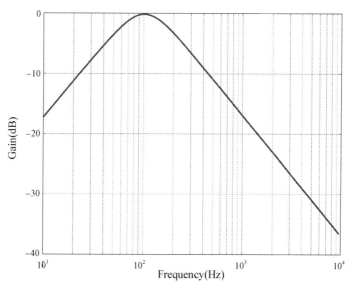

(b) Amplitude Response

Figure 8.11　Second-order Multiple Feedback Bandpass Filter

图 8.11　二阶多反馈带通滤波器

Letting $C_1 = C_2 = C$, by defining a scaling factor $k = \omega_p C$ makes the multiple feedback bandpass filter easy to design. Just following these simple steps:

$$\begin{cases} R_1 = \dfrac{Q}{|A_0| k} \\[2mm] R_3 = \dfrac{Q}{(2Q^2 - |A_0|) \cdot k} \\[2mm] R_2 = \dfrac{2Q}{k} \end{cases} \tag{8.25}$$

8.2.5 Band Stop Filter (带阻滤波器)

The band stop filter (also known as a band-rejection filter) passes all frequencies with the exception of those within a specified stopband which are greatly attenuated. A band stop filter can be represented as a combination of a low pass filter and a high pass filter in parallel if the stopband is wide enough that the two filters do not interact too much.

If the band stop filter has a very narrow stopband, then it is more commonly referred to as a **notch filter**. A notch filter can significantly attenuate specific frequency signals but passes all other frequency components with negligible attenuation. This feature makes the notch filter very practical in order to cancel the unwanted harmonic components (such as 50 Hz/60 Hz power frequency) presented in the input signal.

Filters that having peak or notch at a certain frequency ω_0 are useful to retain or eliminate a particular frequency component of a signal. A generic second-order peak/notch filter is described as

$$A(s) = A_0 \frac{s^2 + g\dfrac{\omega_0}{Q}s + \omega_0^2}{s^2 + \dfrac{\omega_0}{Q}s + \omega_0^2} \tag{8.26}$$

The amplitude response of the peak/notch filter is shown in Figure 8.12. The filter is notch when $|g| < 1$.

A twin-T notch filter and its frequency response are shown in Figure 8.13. It should be a third-order filter, since there are three capacitors in this circuit. However, by choosing $R_1 = R_2 = 2R_3 = R$, $C_1 = C_2 = C_3/2 = C$, the transfer function of the filter is given by

$$A(s) = \frac{s^3 + \dfrac{1}{RC}s^2 + \dfrac{1}{(RC)^2}s + \dfrac{1}{(RC)^3}}{s^3 + \dfrac{3}{RC}s^2 + \dfrac{3}{(RC)^2}s + \dfrac{1}{(RC)^3}} \tag{8.27}$$

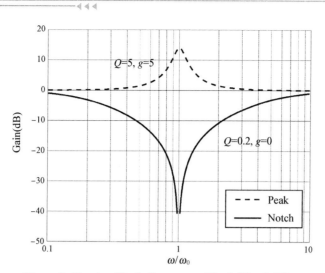

Figure 8.12   Amplitude Response of Peak/Notch Filter

图 8.12   峰/陷波滤波器幅频响应

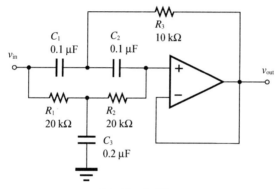

（a）Notch Filter

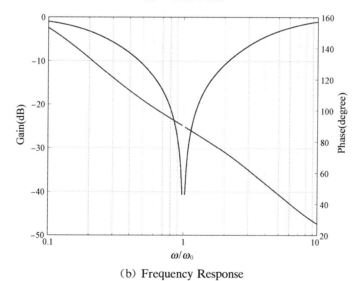

（b）Frequency Response

Figure 8.13   Notch Filter and Its Amplitude Response

图 8.13   陷波器及其频率特性

As the the singularity in $s = -\dfrac{1}{RC}$ is a zero and a pole simultaneously, it can be simplified as a second-order transfer function

$$A(s) = \frac{s^2 + \omega_n^2}{s^2 + \dfrac{\omega_n}{Q}s + \omega_n^2} \tag{8.28}$$

where $\omega_n = \dfrac{1}{RC}$, $Q = \dfrac{1}{2}$.

8.3   Objectives（实验目的）

In this lab, you begin to understand the frequency properties of electronic system. You will be building and investigating several active filters:

1. to find the cutoff frequencies of low pass and high pass filters.
2. to find the central frequency and $-3\,\text{dB}$ bandwidth of a bandpass filter.
3. to measure the voltage gain at different frequencies, and to plot frequency response graph.

8.4   Experimental Procedure（实验步骤）

1. Build a low pass filter and investigate its characteristics.
(1) Refer to Figure 8.6, set filter specification $Q = 0.707$ (Butterworth filter), simulate the low pass filter, plot the filter's frequency response in the frequency range $10\,\text{Hz} - 10\,\text{kHz}$. Listing 8.1 is a netlist as a reference.

Listing 8.1   NGSPICE Netlist for Figure 8.6

清单 8.1   图 8.6 的 NGSPICE 网连表单

```
1    Second Order Low Pass Filter
2    .include op07.mod
3
4    Vin    1    0     dc 0 ac 1
5    R1     1    2     20k
6    R2     2    A     20k
7    C1     2    out   0.2u
8    C2     A    0     0.2u
9    Rp out B 4.1k        ; adjustable
10   R4     B    0     10k
11   X1 A B VCC VEE out OP07
```

```
12  Vcc     VCC     0     12V
13  Vee     VEE     0    - 12V
14
15  .control
16  ac dec 10 10 10k
17  plot db(v(out))
18  plot phase(v(out))
19  .endc
20  .end
```

(2) Construct above low pass filter ($Q=0.707$). Input a bunch of sine waves. Use the channel one of the digital oscilloscope to show the input voltage waveform, and channel two to show the output voltage waveform. Turn on the phase measurement on the oscilloscope, watch the input and output voltage waveforms.

Set the frequency range from 10 Hz to 10 kHz. Measure and plot the magnitude and phase response of the filter. Find the cutoff frequency of the low pass filter.

Refer to equation (8.8), the frequency response at frequency $\omega = \omega_L$ is $A(j\omega_L) = -jA_0 Q$, i. e. the phase response at frequency ω_L is $-90°$. This may help you finding the ω_L (or f_L).

Find the maximum amplitude of the frequency response, then find the $-3\,dB$ point relative to the maximum response. This is the corner frequency f_L of Butterworth prototype. Compare $-3\,dB$ frequency with that of measured by **phase criterion**.

2. Build a second-order high pass filter according to Figure 8.8. Set the Q-factor to 1 by changing R_p. Measure and plot the magnitude response of the filter in the frequency range of 10 Hz to 10 kHz.

Find the high pass cutoff frequency of the filter using phase criterion. According to equation (8.14), the phase response of the high pass filter at $\omega = \omega_H$ is $-90°$.

Compare your measurement curve with the simulation.

3. Refer to Figure 8.10, build a second-order bandpass filter.

- In this circuit, let $R_1 = R_2 = R_3 = R$, $C_1 = C_2 = C$, therefore, $\omega_p = \dfrac{\sqrt{2}}{RC}$,

$$Q = \frac{\sqrt{2}}{4 - A_{VF}}, \quad A_0 = \frac{A_{VF}}{4 - A_{VF}}.$$

- Set Q-factor to 1 by adjusting R_p, measure the central frequency f_p

From equation (8.20), the phase response at $\omega = \omega_p$ is 0. To find the central frequency, use one channel of the digital oscilloscope to show the input voltage waveform while the other channel for the output voltage waveform. Change the frequency until the two sine waves are coincide, or the phase between two waveforms is indicated 0°.

- The -3 dB bandwidth of the filter is give as

$$BW = \frac{\omega_p}{2\pi Q} = \frac{f_p}{Q}$$

The -3 dB points ω_L and ω_H satisfy

$$|A(j\omega)| = 0.707A_0 \qquad (8.29)$$

From equation (8.21) and equation (8.29), the cutoff frequencies at low frequency side and high frequency side can be calculated.

$f_L =$ _____ , $f_H =$ _____ , $BW = f_H - f_L =$ _____

- Measure lower and upper cutoff frequency of the bandpass filter using magnitude criterion.
- Measure and plot amplitude response of the bandpass filter in the range of 100 Hz to 10 kHz. Find the -3 dB points on your graph.
- Compare your measurement with the mathematical analysis.

4. Refer to Figure 8.13 to build a notch filter. Measure and plot its amplitude response. Find the central frequency of the notch filter.

If you want to increase Q-factor of the notch filter, how would you modify the circuit?

8.5 Question(问题)

The Sallen-Key filter has always a limited voltage gain A_0. But what will happen if we adjust R_p in equation(8.10) to set a large $A_0(A_0 \geqslant 3)$ intentionally? Will the circuit still be a low pass filter? If not, what is it?

[实验简介]

滤波器的应用及其广泛.经典的频率滤波器可以去除无用的频率成分,保留有用的部分.在模拟信号领域,滤波器特性通常由电路中的电阻和电容决定,也称阻容滤波器.

使用萨兰-基结构可以构造二阶基本节滤波器.本实验重点研究这类滤波器的构造和特性.滤波器每增加一阶,低通或高通的截止特性衰减率增加 6 dB/倍频程.当对滤波器截止特性要求较高时,可通过多个基本节级联而成.

[Vocabulary]

active filter：有源滤波器

modem：调制/解调（modulation-demodulation 的并称）

attenuation：衰减

anti-aliasing low pass filters：抗混叠低通滤波器

passband：通带

stopband：阻带

causality constraints：因果律约束

passband ripple：通带纹波

transition region：（通带与阻带之间的）过渡带

denominator：分母

numerator：分子

$-6\,dB/octave$：每倍频程$-6\,dB$

decade：十倍频程

rolloff：滚降，增益随频率变化的衰减

eigenfrequency：本征频率

band stop/band-rejection filter：带阻滤波器

order：滤波器阶数

quality factor（Q-factor）：品质因数（Q 值）

notch filter：陷波器

power frequency：工频（市供交流电频率 $50\,Hz$ 或 $60\,Hz$）

[Translations]

1. Filters can be used to separate signals，passing those of interest，and attenuating the unwanted frequencies.

   滤波器可用于分离信号,允许感兴趣的频率通过,同时阻止那些不想要的频率成分.

2. In addition，there are filters that do not filter any frequencies of a complex input signal，but just add a linear phase shift to each frequency component，thus contributing to a constant time delay. These are called all-pass filters.

   此外,还有一类滤波器,它们不过滤复数输入信号的任何频率成分,而只是在各个频率成分中增加一个线性相移,从而贡献一个延时常数.这类滤波器被称为全通滤波器.

3. Unfortunately，the idealized filters cannot be easily built because of causality constraints. The transition from passband to stopband will not be instantaneous，instead，there will be a transition region. Also，the

amplitude response in stopband will not be zero.

不幸的是，由于受到因果律的限制，理想滤波器不容易实现．由通带到止带不会是陡变的，而是存在一个过渡带．另外，止带的幅度响应也不会是 0．

4. The steepness of a filter is affected by the order of the filter. Order is also the number of poles in the transfer function. A pole is a root of the denominator of the transfer function. A pole frequency corresponds to a corner frequency at which the slope of the magnitude curve decreases by -6 dB/octave, or -20 dB/decade. Conversely, a zero is a root of the numerator of the transfer function.

滤波器（截止特性）的陡峭程度受滤波器阶数的影响．阶数也是传递函数中极点的数目．极点是传递函数分母（多项式）的根．极点频率对应转折频率，在这个频率上形成幅度响应曲线每倍频程 -6 dB 或十倍频程 -20 dB 的斜率．对应地，零点是传递函数分子（多项式）的根．

5. Filters only consist of passive components such as inductors, resistors and capacitors are called passive filters or LRC filters. Active filters are circuits that use an operational amplifier as the active device in combination with some resistors and capacitors to provide an LRC-like filter performance.

仅含有被动元件（如电感、电阻和电容）的滤波器称为无源滤波器，或称 LRC 滤波器．有源滤波器是使用运算放大器作为主动设备、组合电阻和电容以提供 LRC 类型滤波器性能的滤波器．

6. By choosing different passive components (e.g. resistors and capacitors) for Z_1, Z_2, Z_3 and Z_4, the filter can be made with low pass, bandpass, and high pass characteristics. Recall that a capacitor with capacitance C has impedance $Z_C = 1/sC$, where $s = \sigma + j\omega$ is the complex frequency variable. For pure sine waves, the damping constant σ becomes zero and $s = j\omega = j2\pi f$.

通过对 Z_1、Z_2、Z_3 和 Z_4 选取不同的被动元件（电阻和电容），滤波器可被构造成低通、带通及高通特性．记得电容值为 C 的容抗是 $Z_C = 1/sC$，这里 $s = \sigma + j\omega$ 是（拉普拉斯域的）复数频率变量．对于纯正弦波，衰减常数 σ 变为 0，$s = j\omega = j2\pi f$．

7. The Q-factor determines the height and width of the peak of the frequency response of the filter. For example, a Butterworth filter (also known as the maximally flat magnitude filter), which has maximally flat passband frequency response, has a Q-factor of $1/\sqrt{2}$.

Q 值决定了滤波器频率响应峰值处的高度和宽度．例如，Butterworth 滤波器（又被称为最大平坦型滤波器），它有着最平坦的通带频率响应，它的 Q 值是 $1/\sqrt{2}$．

8. A notch filter is a band-rejection filter that significantly attenuates specific frequency signals but passes all other frequency components with negligible

attenuation. This feature makes the notch filter very practical in order to cancel the unwanted harmonic components (such as 50 Hz power frequency) presented in the input signal.

陷波器是一种带阻滤波器,它(可以)显著地衰减特定频率的信号,且能让其他频率成分几乎不衰减地通过.这个特性使得陷波器在用于消除输入信号中不需要的谐波成分(如 50 Hz 工频)的场合极具应用价值.

9

Electronic Oscillators
振荡器/波形发生器

An electronic oscillator produces a periodic, oscillating electronic signal, such as sine wave, square wave, triangle wave, etc. A linear oscillator circuit which uses an RC network, a combination of resistors and capacitors, for its frequency selective part is called an RC oscillator. Oscillators convert direct current (DC) from a power supply to an alternating current (AC) signal. They are widely used in many electronic devices ranging from simplest clock generators to digital instruments (like calculators) and complex computers and peripherals etc.

9.1 Rectangular Wave Generator (矩形波发生器)

A square wave generator is also called an astable multivibrator. Square wave and rectangular wave, as their names might suggest, are related, but square and rectangular waves are also distinct waveforms because of their **duty cycles**. A duty cycle is the percentage of the ratio of pulse duration, or pulse width to the total period of the waveform. A rectangular wave which has a duty cycle of 50% is often called square wave.

9.1.1 Square Wave Generator (方波发生器)

A basic square wave generator is shown in Figure 9.1(a). Series circuit with R_1 and R_2 forms a voltage divider at node B, with division factor $\beta = \dfrac{R_1}{R_1 + R_2}$. The

▸▸▸

output voltage V_{out} is shunted to ground by two zener diodes[①] connected back-to-back in series and is railed to $\pm V_Z$. Assume the capacitor C is fully discharged at first and the output of the op-amp is saturated at positive V_Z.

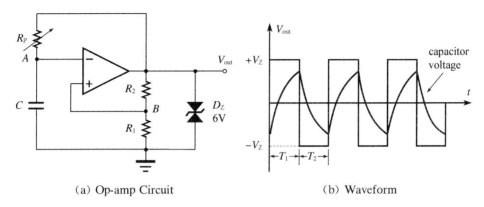

(a) Op-amp Circuit   (b) Waveform

Figure 9.1   Square Wave Generator
图 9.1   方波发生器

The capacitor C starts to charge up from the output voltage $+ V_Z$ through resistor R_p at a rate determined by their RC time constant. A fraction of the output voltage $+ V_Z$ is fed back to the non-inverting terminal by the resistor network R_1 and R_2, which is $\beta \cdot V_Z$. During this charging period, the voltage at node A becomes greater than $\beta \cdot V_Z$ at some point. At this point, the voltage at inverting terminal will be greater than the voltage at the non-inverting terminal of the op-amp and the output swings to the negative rail $- V_Z$. Now the capacitor discharges towards $- V_Z$. At some point, the voltage at node A becomes lower than $- \beta \cdot V_Z$, the output voltage of the op-amp flips again. This cycle is repeated over time and the result is a square wave swinging between $+ V_Z$ and $- V_Z$ at the output of the op-amp.

Let $V_C(t)$ be the voltage across the capacitor C. During charging period,

$$V_C(t) = V_Z + (-\beta V_Z - V_Z)\exp\left(-\frac{t}{R_p C}\right) \tag{9.1}$$

As time $t = T_1$, the voltage across the capacitor C is charged to βV_Z. At this point, the voltage of the inverting terminal will be greater than the non-inverting terminal, thus flips the output of the op-amp.

$$\beta V_Z = V_Z + (-\beta V_Z - V_Z)\exp\left(-\frac{T_1}{R_p C}\right) \tag{9.2}$$

Upon solving, we get

$$T_1 = R_p C \ln\left(\frac{1+\beta}{1-\beta}\right) \tag{9.3}$$

Charging and discharging processes are symmetric, therefore, $T_1 = T_2$. The time period of the square wave generated at the output is:

$$T = 2R_p C \ln\left(\frac{1+\beta}{1-\beta}\right) \tag{9.4}$$

We can see from the above equation that the oscillation frequency depends upon the RC time constant and the feedback fraction β. Set $R_1 = R_2 = 10$ kΩ. By changing R_p, the circuit can generator square wave of different frequencies:

$$f \approx \frac{1}{2.2 R_p C} \tag{9.5}$$

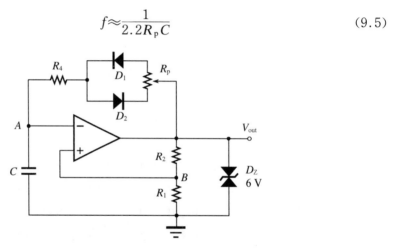

Figure 9.2 Rectangular Wave Generator

图9.2 矩形波发生器

9.1.2 Rectangular Wave with Adjustable Duty Cycle (可调占空比的矩形波)

A type of rectangular wave generator with adjustable duty cycle is shown in Figure 9.2. By adding diodes in series with the charge resistor, the circuit has different charging and discharging loop. The potentiometer R_p is divided into two parts, R_{pp} and R_{pn}. During charging period T_H, the diode D_1 is on and D_2 is off, the upper part of R_p and R_4 forms charging resistor, gives

$$T_H = (R_4 + R_{pp}) C \ln\left(\frac{1+\beta}{1-\beta}\right) \tag{9.6}$$

The discharge cycle given by C, R_4 and R_{pn} loop is

$$T_L = (R_4 + R_{pn}) C \ln\left(\frac{1+\beta}{1-\beta}\right) \tag{9.7}$$

therefore, the oscillator has the period

$$T = T_{\mathrm{H}} + T_{\mathrm{L}} = (2R_4 + R_{\mathrm{p}}) C \ln\left(\frac{1+\beta}{1-\beta}\right) \tag{9.8}$$

and duty cycle

$$q = \frac{T_{\mathrm{H}}}{T} \times 100\% = \frac{R_4 + R_{\mathrm{pp}}}{2R_4 + R_{\mathrm{p}}} \tag{9.9}$$

9.2　Triangular Wave Generator (三角波发生器)

Triangular waves are a periodic, non-sinusoidal waveform with a triangular shape. Triangle and sawtooth waves are somewhat different. A triangular wave has equal rising and falling time while a sawtooth wave has different rising and falling time.

There are many methods for generating triangular waves. Basically, triangular wave is generated by alternatively charging and discharging a capacitor with a constant current. We have learned that the output of integrator is a triangular wave if its input is a square wave. This means that a triangular wave generator can be formed by simply connecting an integrator to the square wave generator as shown in Figure 9.3.

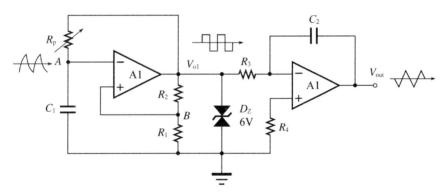

Figure 9.3　Triangular Wave Generator
图 9.3　三角波发生器

The first op-amp outputs a square wave. Due to the zener diode, the amplitude of the output voltage is limited to V_{Z}. The square wave is applied to the inverting input of the second op-amp through the input resistor R_3. Recall that V_{out} is proportional to integration of V_{o1} respect to time, i.e.

$$V_{\mathrm{out}} = -\frac{1}{R_3 C_2} \int V_{\mathrm{o1}}(t) \mathrm{d}t \tag{9.10}$$

The period T of the triangular wave, which is also the same as that of the square wave, is determined by equation (9.4). The amplitude of the triangular wave is given as

$$V_m = -\frac{1}{R_3 C_2}\int_0^{T/2} V_Z dt$$

$$= -\frac{1}{R_3 C_2}\frac{TV_Z}{2} \tag{9.11}$$

As the period increases, the amplitude of the triangular wave also increases linearly. When the frequency is too low, the waveform will be saturated and cause clipping.

Another triangular wave generator, which requires fewer components, is shown in Figure 9.4.

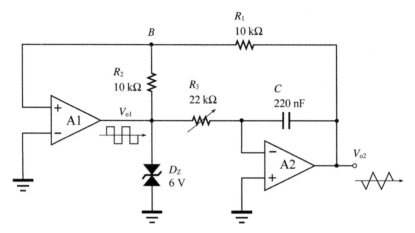

Figure 9.4 Another Triangular Wave Generator
图9.4 另一种三角波发生器

There are two parts of a triangular wave generator circuit. The first part generates a square wave, and the second part converts the square wave into a triangular waveform. The first circuit acts as a comparator which consists of an op-amp and a voltage divider (R_1 and R_2) connected to the op-amp's non-inverting terminal. The output of this comparator acts as input for the second part, which is an integrator circuit. Let's say the first output is V_{o1} and the second output is V_{o2}. V_{o2} is connected with the first op-amp as feedback. The waveform analysis is shown in Figure 9.5.

The comparator continuously compares the voltage at node B (non-inverting terminal) with the inverting terminal, i.e. ground voltage. The output V_{o1} is either $+V_Z$ or $-V_Z$, which is railed by the zener diode, depending on whether the V_B is positive or negative. When the voltage at node B is positive, the comparator gives $+V_Z$ as output.

This output provides input for the second op-amp that produces a negative-

◀◀◀

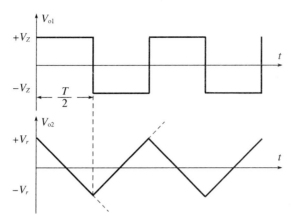

Figure 9.5　Waveform Analysis to Figure 9.4
图9.5　图9.4电路的波形分析

going ramp voltage V_{o2} as output:

$$V_{o2}(t) = V_o - \frac{1}{R_3 C}\int_0^t V_{o1}(\tau)\mathrm{d}\tau = V_o - \frac{V_Z}{R_3 C}t \qquad (9.12)$$

V_{o2} gives negative voltage up to a certain value. After some time, the voltage at B falls below zero, and the comparator gives $-V_Z$ as output. At this point, the value of V_{o2} starts increasing towards the positive voltage. When the value crosses a certain positive voltage, the output of the square wave goes up to $+V_Z$. This process goes on continuously, providing the square wave at V_{o1} as well as the triangular wave at V_{o2}.

For this entire circuit, we notice that when V_{o2} gets changed from positive to negative at the voltage of V_r, a positive saturation voltage at V_{o1} is reached. Similarly, when V_{o2} gets changed from negative to positive at the voltage of $-V_r$, a negative saturation voltage is developed. Resistor R_2 is connected to V_{o1} while resistor R_1 is connected to V_{o2}. As there is no current flows into the non-inverting input of op-amp A1, the voltage at node B can be written as

$$V_B = \frac{R_2}{R_1 + R_2}(V_{o2} - V_{o1}) + V_{o1} = \frac{R_1 V_{o1} + R_2 V_{o2}}{R_1 + R_2} \qquad (9.13)$$

The output of the comparator flips when $V_B = 0$ at $V_{o2} = -V_r$, therefore, the equation can be written as

$$R_1 V_Z - R_2 V_r = 0$$

gives

$$V_r = \frac{R_1}{R_2}V_Z$$

The peak to peak output voltage $V_{PP} = 2V_r = \dfrac{2R_1}{R_2} V_Z$.

Output at the integrator circuit is given by

$$V_o = -\frac{1}{R_3 C} \int_0^t V_{o1} \, d\tau \tag{9.14}$$

here $V_o = V_{PP}$ and $V_{o1} = -V_Z$.

So, by putting the values we get

$$V_{PP} = -\frac{1}{R_3 C} \int_0^{T/2} (-V_Z) \, dt = \frac{V_Z}{R_3 C} \times \frac{T}{2} \tag{9.15}$$

therefore, the period of the output is

$$T = \frac{4R_1 R_3 C}{R_2} \tag{9.16}$$

So, the frequency is

$$f = \frac{R_2}{4R_1 R_3 C} \tag{9.17}$$

9.3   Sine Wave Generator (正弦波发生器)

There are many types of sine wave oscillator circuits and variants. One of the most common gain-stabilized circuits is the Wien bridge oscillator.[1] Figure 9.6 shows schematic of a Wien bridge oscillator that uses diodes to control amplitude. In this circuit, two RC circuits are used, one with the RC components in series and one with the RC components in parallel. The oscillator can also be viewed as a positive gain amplifier combined with a band pass filter that provides positive feedback. This positive feedback forms a very selective second-order frequency dependant band pass filter with a high Q-factor at the resonant frequency f_r. The feedback factor at frequency f_r is

$$F = \frac{V_p}{V_{out}} = \frac{R_1 \parallel X_{C1}}{R_1 \parallel X_{C1} + R_2 + X_{C2}}$$

where $X_C = 1/j2\pi f_r C$.

Let $R_1 = R_2 = R$, $C_1 = C_2 = C$, therefore the resonant frequency $f_r = 1/2\pi RC$, feedback factor $F = 1/3$. At this frequency, the phase difference between the input and output equals zero and the magnitude of the output voltage is at its maximum.

---

[1]   Max Karl Werner Wien (Dec. 25, 1866 – Feb. 22, 1938), German physicist. He developed a type of bridge circuit in 1891 which is used for precision measurement of capacitance. The circuit was later named after him.

R_3, R_p and R_4 in parallel with diodes form a negative feedback loop. This forms a non-inverting amplifier, with the voltage gain of

$$A_{VF} = 1 + \frac{R_p + R_{eq}}{R_3}$$

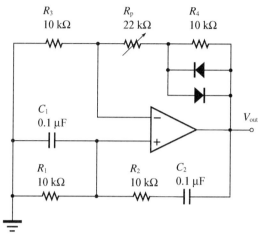

Figure 9.6   Wien Bridge Circuit Schematic

图9.6   文氏桥振荡器电路

where R_{eq} is the equivalent resistor of R_4 in parallel with diodes.

For a linear circuit to oscillate, it must meet the Barkhausen[1] conditions: its loop gain must be one ($|A_{VF} \cdot F| = 1$) and the phase around the loop must be an integer multiple of 2π. In practice, the loop gain is initially larger than one ($|A_{VF} \cdot F| > 1$). As random noise is present in all circuits, and some of that noise will be near the desired frequency. A loop gain greater than one allows the amplitude of frequency to increase rapidly and the circuit will start oscillating.

Once the circuit oscillates, the output voltage increases to force the diode on, thereby reducing R_{eq} and eventually decreasing A_{VF} to $1/3$.

To describe how "pure" a sine wave is, the concept *distortion factor* is introduced. Distortion factor is a measurement of the harmonic distortion present in a signal and is defined as the ratio of the sum of the power of all harmonic components to the power of the fundamental waveform. In practical, THD (*total harmonic distortion*) is most commonly used and defined as the ratio of the RMS amplitude of a set of higher harmonic frequencies V_i to the RMS amplitude of the fundamental or first harmonic sine wave voltage V_1.

$$THD = \frac{\sqrt{V_2^2 + V_3^2 + V_4^2 + \cdots}}{V_1} \tag{9.18}$$

---

[1]   Heinrich Georg Barkhausen (Dec. 2, 1881 – Feb. 20, 1956), German physicist.

9.4 Objectives (实验目的)

The objectives of this lab are:

1. to build and to investigate rectangular wave and triangular wave generator using op-amp.

2. to build and to investigate sine wave generator.

9.5 Experimental Procedure (实验步骤)

9.5.1 Rectangular Wave Generator (矩形波发生器)

1. Build square wave generator according to Figure 9.1(a). Choose resistors $R_1 = R_2 = 10\,\text{k}\Omega$, capacitor $C = 0.1\,\mu\text{F}$, and potentiometer $R_\text{p} = 100\,\text{k}\Omega$. Set the voltage supply to $\pm 12\text{V}$.

Adjust R_p to different values, measure the output voltage and period, fill the recorded data in Table 9.1.

Table 9.1　Square Wave Generator
表 9.1　方波发生器

R_p(kΩ)	Amplitude (V)	T (ms) (measured)	T (ms) (calculated)
20			
40			
60			
80			
100			

2. Refer to Listing 9.1, create your netlist to simulate the circuit using NGSPICE and compare the result with your measurement.

Listing 9.1　Netlist of Squarewave Oscillator, oscillator. cir
清单 9.1　方波振荡器的网连表单文件 oscillator. cir

```
1  OP-Amp Squarewave Oscillator
2  .include op07.mod
3  .model vref6v zener(V_breakdown= 6V I_breakdown= 20 mA
4  +          R_breakdown= 1 I_rev= 1uA I_sat= 1pA)
5
6  Rp      A    out    40k    ; adjustable
7  C       A    0      0.1u
```

```
8   X1        B     A        VCC   VEE   out   OP07
9   R2        out   B        10k
10  R1        B     0        10k
11
12  Az1       out   X        vref6v
13  Az2       0     X        vref6v
14
15  Vcc ·     VCC   0        12V
16  Vee       VEE   0        - 12V
17
18  .control
19  set   color0= white
20  set   color1= black
21  tran      50us  60ms  20ms
22  run
23  plot  v(out)
24  .endc
25  .end
```

Also calculate the theoretical period from equation (9.4). Compare your results. Measure the 2nd, 3rd, 4th and 5th harmonic frequencies of the output using FFT function of the oscilloscope. Theoretically, Square waves with exact 50% duty cycle have no even harmonics.

Using FFT command of NGSPICE, you may also perform harmonic analysis based on Listing 9.1:

```
ngspice 1 -> source oscillator. cir
ngspice 2 -> tran 10us 200ms uic
ngspice 3 -> fft v(out)
ngspice 4 -> plot mag(v(out)) xlimit 0 5k
```

Command "tran" specifies timestep (or sampling interval) of 10 μs, which corresponds to the highest analysis frequency of 50 kHz, and analysis time of 200 ms, which corresponds to the highest frequency resolution of 5 Hz.

3. Build rectangular wave generator with adjustable duty cycle according to Figure 9.2. Choose resistors $R_1 = R_2 = R_4 = 10$ kΩ, capacitor $C = 0.1$ μF, 1N4148 for diodes D_1 and D_2, potentiometer $R_p = 100$ kΩ.

Set potentiometer to about 10 kΩ, 50 kΩ and 90 kΩ respectively, measure the output period and voltage and fill your recorded data in Table 9.2. Compare your measurements with the calculated values.

Fundamental Electrical Experiments

Table 9.2   Rectangular Wave Generator(dc: duty cycle)

表 9.2   矩形波发生器(dc:占定比)

$R_p(\text{k}\Omega)$	$V_{PP}(\text{V})$	Measured			Calculated		
		T (ms)	T_H(ms)	dc(%)	T (ms)	T_H(ms)	dc(%)
10							
50							
90							

9.5.2   Triangular Wave Generator (三角波发生器)

Build triangular wave generator according to Figure 9.4. Change resistor R_3 at different values to generate waveform of different frequencies. Fill the recorded data in Table 9.3.

Table 9.3   Triangular Wave Generator

表 9.3   三角波发生器

$R_3(\text{k}\Omega)$	Measured		Calculated	
	$V_{PP}(\text{V})$	T (ms)	$V_{PP}(\text{V})$	T(ms)
5				
10				
15				

9.5.3   Sine Wave Generator (正弦波发生器)

Build sine wave generator according to Figure 9.6. Adjust the potentiometer R_p to get sine wave with different amplitude. Measure the output amplitude V_o and frequency f, fill the data in Table 9.4. Find the resistance value which the circuit starts oscillating.

Use the FFT function of oscilloscope to measure the harmonics of the output waveform. Let ΔL_n be the amplitude of the n-th harmonic relative to the fundamental frequency. Fill your measurements in Table 9.4 and calculate the distortion. Less distortion means better quality of sine wave output.

Table 9.4   Harmonics of Sinewave Output

表 9.4   正弦波输出的谐波

$R_p(\text{k}\Omega)$	$V_{PP}(\text{V})$	f (Hz)	ΔL_2(dB)	ΔL_3(dB)	ΔL_4(dB)	ΔL_5(dB)	THD (%)
	3						
	6						
	9						
	12						

Circuit starts oscillating when $R_p = $ _____ $k\Omega$

［实验简介］

信号发生器是用于产生重复或单次特定波形信号的电子设备，主要用于设计、测试、维修电子仪器等工作场景，也包括数字计算机中的定时装置.现代电子设备常需要多种类型的测试信号，由此有了方波、三角波、正弦波以及其他特定波波形产生的需求.本实验研究这几种波形的模拟电路生成方法.

进入数字化时代以来，以数字电路为主的信号发生器更为普遍.它产生的信号更加准确，产生信号的类型也更为灵活.如今实验室使用的函数信号发生器都是通过电子计算机产生的数字波形经处理后产生的.

［Vocabulary］

peripheral：外围设备

multivibrator：多谐振荡器

sawtooth wave：锯齿波

zener diode：齐纳二极管

duty cycle：占空比

charge：充电

discharge：放电

total harmonic distortion：总谐波失真（缩写 THD）

根据傅里叶分析理论，非正弦波的周期波形包含了多次谐波，故而矩形波发生器又称"多谐振荡器".理论上，三角波等其他周期波形也是由多次谐波组成.但矩形波在模拟电路和数字电路中都有重要的应用，因此"多谐振荡器"用于专指方波或矩形波发生器而不用于其他类型的波形发生器.

齐纳二极管是利用二极管在反向电压作用下的齐纳击穿效应制造而成的一种具有稳定电压功能的电子器件，因此又称为"稳压二极管".普通二极管反向击穿会产生很大的电流，从而对二极管造成永久损坏.稳压管经特殊设计，使得它在反向击穿时仍然可以长时间稳定工作而不至于损坏（只要电流不超过它的允许值）.

［Translations］

1. A linear oscillator circuit which uses an *RC* network，a combination of resistors and capacitors，for its frequency selective part is called an *RC* oscillator.
   使用 *RC* 网络，将电阻和电容组合用作其选频部件的线性振荡电路被称为阻容振荡器.

2. Square wave and rectangular wave，as their names might suggest，are related，but square and rectangular waves are also distinct waveforms

because of their duty cycles. A duty cycle is the percentage of the ratio of pulse duration, or pulse width to the total period of the waveform. A rectangular wave which has a duty cycle of 50% is often called square wave.

方波和矩形波,正如它们的名称,是有关系的,但方波和矩形波又因其占空比而有所区别. 占空比是脉冲持续时间或脉冲宽度相对于整个波形周期的比率. 占空比为 50% 的矩形波又常被称为方波.

3. The capacitor C starts to charge up from the output voltage $+V_Z$ through resistor R_p at a rate determined by their RC time constant.

在输出电压 $+V_Z$ 作用下,电容 C 通过电阻 R_p 开始以 RC 时间常数充电.

4. Now the capacitor discharges towards $-V_Z$. At some point, the voltage at node A becomes lower than $-\beta \cdot V_Z$, the output voltage of the op-amp flips again. This cycle is repeated over time and the result is a square wave swinging between $+V_Z$ and $-V_Z$ at the output of the op-amp.

此时电容向 $-V_Z$(方向)放电. 到达某个点时,结点 A 的电压变得比 $-\beta \cdot V_Z$ 低,运放的输出电压再次翻转. 这一周期不断重复,其结果就是在运放输出端产生在 $+V_Z$ 和 $-V_Z$ 间上下摆动的方波.

5. A type of rectangular wave generator with adjustable duty cycle is shown in Figure 9.2. By adding diodes in series with the charge resistor, the circuit has different charging and discharging loop. The potentiometer R_p is divided into two parts, R_{pp} and R_{pn}. During charging period T_H, the diode D_1 is on and D_2 is off, the upper part of R_p and R_4 forms charging resistor.

一种可调占空比的矩形波发生器如图 9.2 所示. 通过添加二极管与充电电阻串联,使电路具有不同的充电和放电回路. 电位器 R_p 被分成 R_{pp} 和 R_{pn} 两部分. 在充电周期 T_H,二极管 D_1 导通、D_2 截止,R_p 上部与 R_4 构成充电电阻.

6. Triangular waves are a periodic, non-sinusoidal waveform with a triangular shape. Triangle and sawtooth waves are somewhat different. A triangular wave has equal rising and falling time while a sawtooth wave has different rising and falling time.

三角波是三角形状的周期、非正弦波形. 三角波和锯齿波有点不同. 三角波(波形)具有相等的上升和下降时间,而锯齿波具有不同的上升和下降时间.

7. The first circuit acts as a comparator which consists of an op-amp and a voltage divider connected to the op-amp's non-inverting terminal. The output of this comparator acts as input for the second part, which is an integrator circuit.

电路第一部分是由一个运放和接在运放同相输入端的分压器组成的比较器. 该比较器的输出作为第二部分的输入,第二部分是一个积分电路.

8. The comparator continuously compares the voltage at node B (non-inverting

terminal) with the inverting terminal, i. e. ground voltage. The output V_{o1} is either $+V_Z$ or $-V_Z$, which is railed by the zener diode, depending on whether the V_B is positive or negative. When the voltage at node B is positive, the comparator gives $+V_Z$ as output. This output provides input for the second op-amp that produces a negative-going ramp voltage V_{o2} as output.

比较器持续将结点 B（也就是同相输入端）的电压与反相输入端（也就是地电压）进行比较.输出端 V_{o1} 受齐纳二极管限制,要么是 $+V_Z$,要么是 $-V_Z$,依 V_B 的正负而定.当结点 B 的电压为正时,比较器输出 $+V_Z$.该输出提供第二个运放的输入,产生负斜向变化的电压 V_{o2} 作为其输出.

9. The oscillator can also be viewed as a positive gain amplifier combined with a band pass filter that provides positive feedback. This positive feedback forms a selective second-order frequency dependant band pass filter with a high Q-factor at the resonant frequency f_r.

此振荡器亦可视为一个正增益放大器与带有正反馈的带通滤波器的组合.这个正反馈形成了具有选频特性的、与频率相关的、在谐振频率 f_r 处具有高 Q 值的二阶带通滤波器.

10. For a linear circuit to oscillate, it must meet the Barkhausen conditions: its loop gain must be one ($|A_{VF} \cdot F| = 1$) and the phase around the loop must be an integer multiple of 2π. In practice, the loop gain is initially larger than one ($|A_{VF} \cdot F| > 1$). As random noise is present in all circuits, and some of that noise will be near the desired frequency. A loop gain greater than one allows the amplitude of frequency to increase rapidly and the circuit will start oscillating.

要一个线性电路产生振荡,必须满足 Barkhausen 条件:其回路增益必须为 1 ($|A_{VF} \cdot F| = 1$),且回路的相位必须是 2π 的整倍数.实用中,回路增益开始时大于 1($|A_{VF} \cdot F| > 1$).由于整个电路中存在随机噪声,有噪声在希望产生的频率附近.回路增益大于 1 允许该频率信号的幅度迅速被放大,从而电路起振.

10

Signal Decomposition and Synthesis
信号的分解与合成

According to the theory of **Fourier analysis**，any periodic signal such as square wave is composed of a series of sine waves，therefore a group of filters can extract each sine wave from original signal，with certain phase shift. On the other hand，a square wave may be approximated by summing of a group of sine waves，as long as their frequencies，amplitudes and phases meet certain conditions.

In this lab，you will be building a circuit which includes signal generator，active filters，phase shifters and signal synthesizer. After this experiment，you will have a better understanding of Fourier analysis.

10.1　System Composition（系统组成结构）

The circuit includes four parts and is shown in Figure 10.1.

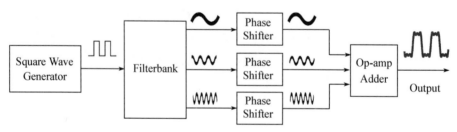

Figure 10.1　Signal Decomposition and Synthesis
图 10.1　信号分解与合成

First of all，you need to build a square wave generator to generate a required square wave with certain frequency and amplitude. Secondly，several filters are used to extract harmonics from the square wave. You may use either low pass filter or band pass filter to extract the fundamental sinc wave，and band pass filters to get the third，the fifth or even higher harmonics（question：why don't we care about the second and the fourth harmonics?). In the third part, phase shifters are applied to each harmonics to adjust the phase of the waveform. You can take one of the harmonic waveform as a reference, therefore omit the phase shifter for this channel. Lastly, phase shifted waveforms will be added proportionally to rebuild an approximated square wave.

Phase shifters are required in this circuit. Phase shifters are used to change the transmission phase angle of an input signal. Ideally, a phase shifter provides an output signal with an equal amplitude to the input signal. The input signal is shifted in phase at the output based on the phase shift provided by the selected phase shifter. Phase shifters are also called all-pass filters. It is called all-pass because it maintains a constant gain for all the frequencies within the operating range.

Figure 10.2 shows two types of first-order all-pass filter, one for lagging phase angles and one for leading phase angles.

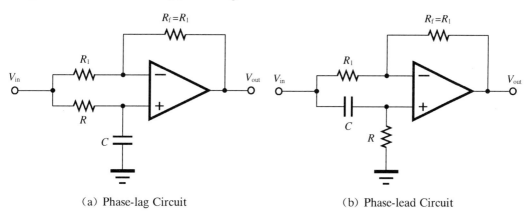

(a) Phase-lag Circuit   (b) Phase-lead Circuit

Figure 10.2   Two Types of All-pass Filter

图 10.2   两种全通滤波器

To analyze the circuit operation, take Figure 10.2(a) as an example, it is assumed that the inverting terminal and non-inverting terminal have the same voltage, i.e. $v_+ = v_-$, and no current flows into the op-amp, gives

$$\frac{X_C}{R + X_C} v_{in} = \frac{R_1 v_{out} + R_f v_{in}}{R_f + R_1} = \frac{1}{2} (v_{in} + v_{out}) \tag{10.1}$$

thus,

$$A(j\omega) = \frac{v_{out}}{v_{in}} = \frac{X_C - R}{X_C + R} = \frac{1 - j\omega RC}{1 + j\omega RC} \tag{10.2}$$

where X_C is the impedance of the capacitor. $X_C = 1/j\omega C$.

Since the numerator and denominator are complex conjugates, their magnitudes are identical [thus $|A(j\omega)| = 1$] and the overall phase angle equals the difference between the angle of numerator and the angle of denominator.

$$\angle A(\mathrm{j}\omega) = -2\tan^{-1}(\omega RC) \qquad (10.3)$$

Using the similar analysis method, you may formulate the phase response of Figure 10.2(b) as

$$\angle A(\mathrm{j}\omega) = \pi - 2\tan^{-1}(\omega RC) \qquad (10.4)$$

10.2 Fourier Analysis of Square Wave (方波的傅里叶分析)

In mathematics, Fourier analysis is the study of how general functions can be decomposed into trigonometric or exponential functions with definite frequencies.

According to Fourier analysis theory, **Fourier series** of a square wave $s(t)$ with angular frequency ω and amplitude A can be written as

$$s(t) = \frac{4A}{\pi} \sum_{k=0}^{\infty} \frac{1}{2k+1} \sin[(2k+1)\omega t] \qquad (10.5)$$

We can see from equation (10.5) that, for the periodic, bipolar, 50% duty cycle square wave, only odd harmonics are present in this waveform. The higher the harmonic frequency is, the smaller the amplitude of this harmonic component. When considering only the first three components, equation (10.5) gives the following approximation

$$s(t) \approx A_o \left[\sin(\omega t) + \frac{1}{3}\sin(3\omega t) + \frac{1}{5}\sin(5\omega t) \right]$$

More harmonics will make up the output square wave more accurate.

10.3 Design and Evaluation (设计与评估)

1. Build your circuit on a breadboard. Build the square wave generator with frequency of 1 kHz. Build at least two filters to obtain signals of fundamental frequency and the third harmonic. When possible, build the third filter to obtain the fifth harmonic.

2. To evaluate the quality of square wave generator, measure the second, third, fourth and fifth harmonics of the square wave using FFT function of the oscilloscope. Idealized square waves have no even harmonics, and the amplitude ratio of fundamental, third and fifth harmonic should be $1 : \frac{1}{3} : \frac{1}{5}$.

3. Measure the total harmonic distortion of each filter to evaluate its performance.

◀◀◀

4. Measure the phase shift of each filter at the given frequency. Build a phase shifter for either fundamental sine wave or the third harmonic channel based on the measured phase shifts. You need to choose a correct phase shifter type（phase-lag circuit or phase-lead circuit）with appropriate *RC* values.

5. Use software tools to plot magnitude and phase response of the filters. Compare the theoretical analysis with your measurements.

6. Build an op-amp adder to sum the two channel signals with correct phase and amplitude ratio.

7. When possible，build the third filter to extract the fifth harmonic from the square wave. Build phase shifter for this channel and add the phase-shifted output to the adder with correct amplitude to form a better square wave output.

8. Measure the amplitudes of each harmonic component of the synthesized waveform.

Challenging experiment：

Analyze the harmonics of triangular wave. Build a circuit that converts square wave to an approximated triangular wave based on your analysis.

［实验简介］

信号的傅里叶分析理论告诉我们,任何周期信号都可以视为一系列与该周期所对应频率成整倍数的正弦波的叠加.本实验为综合设计实验,利用已经学过的模拟电路知识,完成一个信号分解与合成的电路,初步理解傅里叶级数的意义.

作为设计实验,应先根据设计要求计算电路参数,合理选择元器件,并尽可能进行仿真验证,在此基础上完成电路搭建与测试.测试结果应与设计目标及仿真测试相印证.

［Vocabulary］

decomposition：分解

synthesis：合成

phase shifter：移相器

lagging phase：滞后相位

leading phase：超前相位

trigonometric：三角函数

Fourier series：傅里叶级数

numerator：（数学分式中的）分子

denominator：分母

conjugate：共轭

[Translations]

1. Phase shifters are used to change the transmission phase angle of an input signal. Ideally，a phase shifter provides an output signal with an equal amplitude to the input signal. The input signal is shifted in phase at the output based on the phase shift provided by the selected phase shifter. Phase shifters are also called all-pass filters. It is called all-pass because it maintains a constant gain for all the frequencies within the operating range.

   移相器用于改变一个输入信号的传输相位角.理想情况下,移相器产生与输入信号等幅度的输出信号.输入信号到输出端的相移取决于所选移相器提供的相移.移相器也叫全通滤波器.之所以被叫作全通,是因为它在工作频率范围内,幅度增益保持恒定.

2. Since the numerator and denominator are complex conjugates，their magnitudes are identical [thus $|A(\mathrm{j}\omega)| = 1$] and the overall phase angle equals the difference between the angle of numerator and the angle of denominator.

   由于分子和分母是共轭复数,它们的幅值是一样的[所以$|A(\mathrm{j}\omega)| = 1$],且总相位角等于分子和分母的相位角之差.

3. We can see from equation (10.5) that，for the periodic, bipolar，50% duty cycle square wave，only odd harmonics are present in this waveform. The higher the harmonic frequency is，the smaller the amplitude of this harmonic.

   我们可以从公式(10.5)中看到,对于周期性的、双极性、50%占空比的方波,波形中只存在奇次谐波.谐波频率越高,该谐波的幅度越小.

Bibliography

[1] RIGOL User's Guide, DM3068 Digital Multimeter[R]. RIGOL Technologies, Inc., 2014.

[2] Keysight Technologies,Keysight InfiniiVision 1000X-Series Oscilloscopes User's Guide[R].

[3] RIGOL User's Guide,DG4000 Series Function/Arbitrary Waveform Generator [R]. RIGOL Technologies, Inc., 2015.

[4] Michal, Square Waveform—square wave FFT, Harmonics and Spectrum[EB/OL]. (2022 – 06 – 01)[2022 – 03 – 03]. https://911electronic. com/square-waveform-square-wave-fft-harmonics-and-spectrum/.

[5] Keysight Technologies, The Fundamentals of Signal Analysis[R]. http://literature. cdn. keysight. com/litweb/pdf/5952-8898E. pdf.

[6] Robert A. Witte, Spectrum and Network Measurements (Second Edition), Chapter 4[M]. SciTech Publishing, 2014.

[7] Electronics Tutorials, Kirchhoffs Circuit Law[EB/OL]. (2021 – 05 – 13)[2020 – 06 – 04]. https://www. electronics-tutorials. ws/dccircuits/dcp_4. html.

[8] ScienceDirect, Kirchhoff Law [EB/OL]. (2022 – 03 – 02)[2018 – 04 – 20]. https://www. sciencedirect. com/topics/engineering/kirchhoff-law.

[9] Richard Feynman, Robert Leighton, Matthew Sands, The Feynman Lectures on Physics, Vol. II Chapter 22: AC Circuits [EB/OL]. https://www. feynmanlectures. caltech. edu.

[10] GW Instek, GPD-X303S Series User Manual[R]. Good Will Instrument Co.,Ltd.

[11] Holger Vogt, Marcel Hendrix, Paolo Nenzi,Ngspice User's Manual[EB/OL]. (2021 – 03 – 01)[2019 – 09 – 22]https://ngspice. sourceforge. io/docs. html.

[12] Martin Plonus, Electronics and Communications for Scientists and Engineers (Second Edition), Chapter 1[M]. Elsevier, 2020 (ISBN: 978 – 0 – 12 – 817008 – 3).

[13] ScienceDirect, Thevenin Equivalent Circuit [EB/OL]. (2022 – 03 – 02)[2018 – 04 – 20]. https://www. sciencedirect. com/topics/engineering/ thevenin-equivalent-circuit.

[14] Tony R. Kuphaldt, Lessons In Electric Circuits (Fifth Edition), Chapter 10 [EB/OL]. https://www. allaboutcircuits. com/textbook/direct-current/chpt-

10/thevenin-norton-equivalencies/.

[15] Mitchel E. Schultz, Grob's Basic Electronics (13th Edition), Chapter 29[M]. McGraw-Hill, 2021 (ISBN: 978 - 1 - 260 - 57144 - 8).

[16] Russell L. Meade, Robert Diffenderfer, Foundations of Electronics: Circuits and Devices [M]. Delmar, Cengage Learning, 2007 (ISBN: 978 - 1 - 14180 - 0537 - 5).

[17] Electrical4U, Differential Amplifiers: Gain, OP Amp and BJT Circuit [EB/OL]. (2022 - 03 - 01) [2021 - 03 - 07], https://www. electrical4u. com/differential-amplifier/.

[18] Doug Mercer, Differential amplifiers[EB/OL]. (2022 - 03 - 01) [2017 - 06 - 06], https://wiki. analog. com/university/courses/electronics/text/chapter-12.

[19] Ron Mancini, Understand Basic Analog—Ideal Op Amps[EB/OL]. https://www. ti. com. cn/cn/lit/an/slaa068b/slaa068b. pdf.

[20] Paul Horowitz, Winfield Hill, The Art of Electronics (Third Edition), Chapter 4[M]. Cambridge University Press, 2015. (ISBN: 978 - 0 - 521 - 80926 - 9).

[21] Electronics Tutorials, The Differentiator Amplifier [EB/OL]. (2022 - 03 - 01) https://www. electronics - tutorials. ws/opamp/opamp_7. html.

[22] Electronics Tutorials, The Integrator Amplifier [EB/OL]. (2022 - 03 - 01) [2017 - 06 - 06], https://www. electronics-tutorials. ws/opamp/opamp _ 6. html.

[23] Ravi Teja, Operational Amplifier as Integrator [EB/OL]. (2022 - 03 - 01) [2021 - 04 - 10], https://www. electronicshub. org/operational-amplifier-as-integrator/.

[24] Ravi Teja, Operational Amplifier as differentiator [EB/OL]. (2022 - 03 - 01) [2021 - 04 - 14], https://www. electronicshub. org/operational-amplifier-as-differentiator/.

[25] Bruce Carter, Ron Mancini, Op Amps for Everyone (Fifth Edition), Chapter 16[M]. Elsevier Inc. , 2018 (ISBN: 978 - 0 - 12 - 811648 - 7).

[26] James M. Fiore, Operational Amplifiers and LinearIntegrated Circuits - Theory and Application[EB/OL]. (2022 - 03 - 01) [2021 - 05 - 23]. https://eng. libretexts. org/Bookshelves/Electrical _ Engineering/Electronics/Operational _ Amplifiers_ and _ Linear _ Integrated _ Circuits _-_ Theory _ and _ Application _ (Fiore).